LECTURES ON NUCLEAR THEORY

L. D. Landau
and
Ya. Smorodinsky

DOVER PUBLICATIONS, INC.
New York

This Dover edition, first published in 1993, is an unabridged and unaltered republication of the edition published by Plenum Press, Inc., New York, in 1959.

Library of Congress Cataloging-in-Publication Data

Landau, L. D. (Lev Davidovich), 1908–1968.
  [Lektsii po teorii atomnogo iadra. English]
  Lectures on nuclear theory / L.D. Landau and Ya. Smorodinsky.
      p.   cm.
  Originally published: New York : Plenum Press, 1959.
  ISBN-13: 978-0-486-67513-8 (pbk.)
  ISBN-10: 0-486-67513-0 (pbk.)
  1. Nuclear physics. I. Smorodinskii, IA. A. (IAkov Abramovich) II. Title.
QC173.L2463 1993
539.7—dc20   92-47392
                                                  CIP

Manufactured in the United States by LSC Communications
67513005   2022
www.doverpublications.com

Foreword

This book is based on a series of lectures delivered to experimental physicists by one of the authors (L. Landau) in Moscow in 1954.

In maintaining the lecture form in the printed edition we are emphasizing the fact that the presentation makes no pretense at completeness and that the choice of subject matter is purely arbitrary.

Since there is, at the present time, no rational theory of nuclear forces, we have limited our conclusions concerning nuclear structure to those which can be reached from an analysis of the available experimental data, using only general quantum-mechanical relations.

No attempt has been made to give a bibliography of the literature; rather we have indicated only new experimental results. (A rather complete list of references is given in Blatt and Weisskopf, Theoretical Nuclear Physics, Foreign Literature Press, 1954).*

L. LANDAU AND YA. SMORODINSKY

---

* [Blatt and Weisskopf, Theoretical Nuclear Physics, John Wiley and Sons, Inc., New York, N.Y., 1952.] Reprinted by Dover Publications, Inc., 1991

Contents

		Page
Foreword		**v**
LECTURE ONE: Nuclear Forces		1
LECTURE TWO: Nuclear Forces (Scattering of Nucleons by Nucleons).		13
LECTURE THREE: Nuclear Forces (Scattering of Nucleons at High Energies)		23
LECTURE FOUR: Nuclear Structure (Independent Particle Model)		33
LECTURE FIVE: Structure of the Nucleus (Light Nuclei)		43
LECTURE SIX: Structure of the Nucleus (Heavy Nuclei)		53
LECTURE SEVEN: Nuclear Reactions (Statistical Theory) . . .		69
LECTURE EIGHT: Nuclear Reactions (Optical Model. Deuteron Reactions)		77
LECTURE NINE: π-Mesons		87
LECTURE TEN: Interaction of π-Mesons with Nucleons		99

Nuclear Forces

With the discovery of the neutron it became clear that the structural units of which the atomic nucleus is composed are two kinds of particles: protons and neutrons, usually called nucleons.

Of these particles only the proton is stable. The second particle—the neutron—is unstable. In the free state the neutron is radioactive; it is transformed into a proton, emitting an electron and a neutrino

$$n \rightarrow p + e^- + \nu.$$

The half-life of the neutron is approximately 10 min.

It can be shown that the neutron is basically not an "elementary" particle and that it is more appropriately described, in contrast with the proton, as a compound particle, since the neutron can decay into "simpler" particles. On the other hand, although in the free state the proton is a stable particle, in the bound state (inside the nucleus) it can decay into a neutron, positron and neutrino:

$$p \rightarrow n + e^+ + \nu.$$

On this basis we may consider the proton as a complex particle which is transformed into the "simpler" neutron.

Essentially the foregoing means that the question of which of the two particles is the more elementary is not physically meaningful. Both particles are of equal rank; the question of which particle is more susceptible to decay depends only on energy considerations. The free neutron is heavier than the free proton, hence only one of the reactions—neutron decay—can be realized for *free* nucleons. Within the nucleus both reactions are possible and the nature of the decay is determined by the mass of the decaying nucleus and the masses of the nuclei which are possible decay products.

These properties of nucleons mean that it is valid to consider both as elementary particles which can be transformed into each other. In those phenomena, however, in which β-decay of the nuclei does not take place (because of its small probability), we can ignore the neutron-proton transformation.

We shall start with the characteristics of the proton and neutron. The basic difference between these particles lies in the fact that the neutron, as indicated by its name, is a neutral particle whereas the proton has a positive electric charge, equal in magnitude to that of the electron ($4.80 \cdot 10^{-10}$ cgs units). Both of these particles have the same spin, $1/2$ (in units of $\hbar$, in which this quantity is usually measured). The masses of both particles are almost the same: the

1

mass of the proton is 1836 electron masses ($1.6 \cdot 10^{-24}$ g), and the neutron is 2.5 electron masses (about 1.3 Mev) heavier. In the decay of the neutron 0.51 Mev goes into the formation of the electron and 0.78 Mev into the kinetic energy of the particles which are formed.

This small difference in the masses (less than 0.2%) and the fact that the spins are the same are more than coincidental. If one disregards the difference in electric properties the proton and neutron are very much the same, and this similarity is of fundamental significance in the physics of the nucleus.

This similarity is the basis for many nuclear phenomena and will not be labored. The effect of the similarity is most vividly seen in so-called "mirror nuclei."

If we consider two nuclei, the second of which is the same as the first with all protons replaced by neutrons and all neutrons replaced by protons, we have a pair of nuclei which are known as *mirror* nuclei. Since the number of particles is the same in both nuclei, the atomic weights of both nuclei are also almost the same. If the first nucleus has an atomic number Z, obviously the second has an atomic number $A - Z$ (A is the mass number).

Mirror nuclei of this type are well known in the light mass region (it is apparent that the heavier nucleus must be unstable). The first such pair is comprised of the neutron and proton themselves. Examples of subsequent pairs are He^3, H^3; Be^7, Li^7; B^9, Be^9; C^{14}, O^{14}; and so forth. It has been found experimentally that both mirror nuclei have very similar properties—almost identical binding energies, similar excitation spectra for the excited levels, identical spins. This symmetry in the properties of mirror nuclei is undoubtedly due to the properties of the interaction between protons and neutrons, i. e., the symmetry properties of nuclear forces.

From the foregoing, it is seen that, aside from the relatively weak electric forces which act between two protons, all the available experimental facts indicate that the forces between two protons are very similar to the forces between two neutrons, a feature which is known as the *charge symmetry of nuclear forces*. As we shall see later, the charge symmetry of nuclear forces is actually the manifestation of a still more fundamental characteristic, the so-called *isotopic invariance*.

Contemporary nuclear physics still has no theory of nuclear forces. Hence, the only presently available sources of knowledge on nuclear forces are the experimental results. Thus, in formulating the properties of nuclear forces it is natural to rely on the experiments from which this knowledge is obtained.

The simplest nuclear system is a system composed of a proton and a neutron and the most elementary experiments are those involving scattering of neutrons by protons, and deuteron experiments. The very existence of the deuteron—a compound particle consisting of a proton bound to a neutron (with a binding energy of 2.23 Mev)—indicates that the nuclear interaction is rather strong and that it is, to some degree at least, an essential factor in the attraction between these particles. The latter statement follows because the particles could not come together to form a stable compound particle if only repulsive forces were acting.

It does not follow, however, that this interaction is responsible for the potential energy, 2.23 Mev, as would be the case for two particles at rest in classical mechanics. The classical picture of two bound particles at rest cannot be carried over into quantum mechanics. Actually, according to the very significance of the words "bound system," in such a system the particles are close to each other, and consequently the uncertainty in the mutual position Δr is small; whence, from the laws of quantum mechanics it follows that the momentum cannot be small, and must be at least of order $\hbar/\Delta r$. In particular, this situation means that the binding energy is not determined exclusively by the interaction energy but also by the effective distances over which the nuclear forces must act.

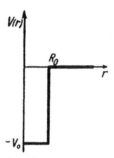

Fig. 1. The potential as a function of the distance between particles in the "square-well" model.

In order to investigate the quantum-mechanical properties of the interaction, we consider a simple model for the proton-neutron interaction—the so-called "square-well potential." In particular, we assume that the potential due to the nuclear forces is zero at distances $r > R_0$, where R_0 is some constant; at smaller distances it is assumed that the potential is constant and equal in absolute value to V_0 (since the nuclear forces are attractive forces this potential is negative). A curve showing the potential $V(r)$ is given in Fig. 1, from which the origin of the designation "square-well" becomes clear.

Recalling that the force is the derivative of the potential, we see that the square-well potential gives rise to a configuration of two particles which interact only when the distance between them is exactly R_0, since the potential $V(r)$ is constant for $r < R_0$; at the point $r = R_0$, however, the interaction force becomes infinite. It is obvious that such a potential is not satisfactory from a physical point of view. However, this simplified model is useful for making calculations and gives a fairly good picture of the basic properties of the quantum-mechanical interaction; for these reasons it is frequently used for illustrative purposes.

First of all we shall consider the effect of increasing the depth V, keeping the radius of the well equal to R_0. At small values of V it turns out that the system can have no stable states. In order for such a state to exist, that is, in order for a system to be formed, the depth of the well must satisfy the inequality

$$V > \frac{\pi^2}{8} \frac{\hbar^2}{\mu R^2},$$

where μ is the reduced mass of the system: $1/\mu = (1/m_1) + (1/m_2)$; if $m_1 = m_2$, for example in a system of two nucleons, $\mu = m/2$. This inequality (aside from the numerical coefficient) follows immediately if one notes that the momentum is of order $\hbar/R$, which, as has already been mentioned, is to be associated with a particle in a potential well of radius R, and leads to a kinetic energy of order $\hbar^2/\mu R^2$. Roughly speaking, this inequality leads to the requirement that the depth of the well must be greater than the kinetic energy of the particle.

When the depth of the well exceeds $(\pi^2/8)(\hbar^2/\mu R^2)$, the system has a single level. With a further increase in the depth of the well the energy of the level increases in absolute value. However, the level energy is always considerably smaller than the depth of the well.

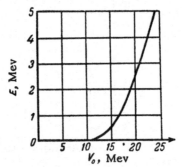

Fig. 2. Energy of the level for a square-well potential. The radius of the well is taken as $2.80 \cdot 10^{-13}$ cm.

The level energy remains relatively small as the depth of the well increases to high values. This situation is illustrated in Fig. 2, which shows the energy of the level as a function of the well depth. It is obvious from this figure that care must be used in estimating the strength of the interaction on the basis of binding energy data because the interaction energy can exceed the binding energy by a large margin. This effect is well illustrated in the deuteron. The bound state of the deuteron lies close to the top of the potential well so that the binding energy of 2.23 Mev is considerably smaller than the interaction energy of the neutron and proton. This property of the deuteron is extremely important in analyses of scattering of neutrons by protons. We now consider this phenomenon, recalling certain features of the quantum theory of scattering which will be required in what follows.

An analysis of the scattering of neutrons by protons is somewhat complicated by the fact that these particles have spin. However, an analysis of scattering of spinless particles reveals the main features of the more complete theory. Thus, we shall use the results of the more simple theory of scattering of spinless particles in a spherically symmetric force field (i. e., a field in which the potential is independent of angle). In the quantum-mechanical formalism scattering in such a force field is described by a superposition of plane waves, corresponding to the incident particle, and spherically scattered waves, representing particles scattered by the force center.

A plane wave, propagating along the positive z-axis, is represented by a wave function of the form e^{ikz} [with an amplitude of unity, this function represents a flux of v particles per second through 1 cm^2; $v = (1/m)\hbar k$ is the particle velocity]. At large distances the spherical wave due to the presence of the scattering center is given by a function of the form $f(\theta)(e^{ikr}/r)$, where r is the distance from the scattering center and $f(\theta)$ is the scattering amplitude; $f(\theta)$ is a function of polar angle, and describes the angular distribution of scattered particles. The positive exponential means that the direction of scattering is from the scattering center toward infinity.

The scattering amplitude introduced in this way is a quantum-mechanical feature of the process. It is easy to show that $f(\theta)$ has the dimensions of length. This follows from the fact that $f(\theta)/r$ must have the same dimensionality as e^{ikz}, that is to say, it must be dimensionless. Hence, $f(\theta)$ is sometimes known as the *scattering length* (usually this designation is applied at small particle energies since $f(\theta)$ becomes a constant which is independent of angle and energy). The experimentally measured scattering cross sections are related to $f(\theta)$ by the simple relation

$$d\sigma = |f(\theta)|^2 do$$

(do is an element of solid angle).

Quantum mechanics furnishes a method for relating the scattering amplitude to the properties of the system. In this analysis it is necessary, first of all, to resolve the incident wave into spherical waves: the incident flux with a given value of particle momentum, $\hbar k$, is given in the form of a sum of fluxes each term of which is characterized by a particular value of angular momentum. This procedure is valid because in a field of central symmetry the angular momentum is an integral of the motion, being described by the azimuthal quantum number L (the square of the angular momentum $|\mathbf{L}|^2 = L(L + 1)\hbar^2$). Each spherical wave with a given L will have its own angular dependence, which is described by the factor $P_L(\cos \vartheta)$, the Legendre polynomial. The wave with $L = 0$ is spherically symmetric since $P_0(\cos \vartheta) = 1$; the wave with $L = 1$ contains the factor $P_1(\cos \vartheta) = \cos \vartheta$, the wave with $L = 2$ contains the factor $P_2(\cos \vartheta) = (3 \cos^2 \vartheta - 1)/2$ and so on.

It is shown in quantum mechanics that in the resolution of a plane wave into spherical waves, at large distances from the origin, the term associated with angular momentum $L\hbar$ is of the form

$$\Psi_L \sim \frac{1}{r} \sin \left[\mathbf{k} \cdot \mathbf{R} + \frac{L\pi}{2} \right] P_L (\cos \vartheta) \qquad (r \rightarrow \infty).$$

If there is a scattering center in the path of the incident beam the form of the function is changed. However, at large distances from the origin the dependence on distance remains virtually unchanged because at these distances the field has almost vanished and the wave function must satisfy the wave equation for a free particle. The scattering process manifests itself in the fact that the phase is changed by an amount δ_L:

$$\Psi_{L'} \sim \frac{1}{r} \sin\left[\mathbf{k}\cdot\mathbf{R} + \frac{L\pi}{2} + \delta_L\right].$$

Thus the change of phase (or more simply, the phase shift), δ_L, is a parameter which describes the scattering of the Lth spherical wave.

The phase shift, δ_L, is connected with the scattering amplitude by a simple relation. Specifically:

$$f(\theta) = \lambdabar \sum_L (2L + 1) \frac{e^{2i\delta_L} - 1}{2i} P_L(\cos\vartheta),$$

where

$$\lambdabar = \frac{1}{k},$$

or, in somewhat different form,

$$f(\theta) = \lambdabar \sum_L (2L + 1)e^{i\delta_L} \sin\delta_L P_L(\cos\vartheta).$$

This expression is complex; to obtain the scattering cross section it is necessary to compute the square of the modulus $f(\theta)$.

It is apparent that the total scattering cross section is given by the expression

$$\sigma_t = \int |f(\vartheta)|^2 d\Omega = 4\pi\lambdabar^2 \sum_L (2L + 1) \sin^2\delta_L.$$

For example, for waves characterized by $L = 0$, we obtain:

$$\sigma(L = 0) = 4\pi\lambdabar^2 \sin^2\delta_0.$$

Scattering of particles with zero angular momentum is called S-scattering.

It is obvious that the cross section for S-scattering cannot be arbitrarily large. Since $\sin\delta_0$ cannot be greater than unity,

$$\sigma_L(L = 0) \leq 4\pi\lambdabar^2.$$

For waves characterized by $L = 1$ (P-scattering), $L = 2$ (D-scattering), etc. the limiting cross section is $4\pi(2L + 1)\lambdabar^2$.

We see that the scattering is completely characterized by the phase shifts. Hence, the first problem in an analysis of the results of an experiment is the determination of the phase shift; the phase value determined in this way yields information on the essential features of the interaction between the particles. Strictly speaking, a knowledge of the scattering phase allows one to calculate the dependence of the interaction force on distance; this calculation, however, requires extremely detailed information on the phase shifts, and has not been carried out at the present time. It should be emphasized that the determination of phase shifts from measured cross sections is in itself a complicated problem; this problem is particularly difficult if the particles have spin. In this case it is necessary to carry out experiments with polarized particles to obtain complete information.

We now consider the possible values which may be assumed by the phase. It is obvious that the phases δ_L and $\delta_L + \pi$ yield the same wave function. Thus, the phase is determined only to within an arbitrary factor, a multiple of π. To eliminate the ambiguity, it is convenient to limit the range of variation of the phase. Thus, it is usually assumed that either

$$0 \leq \delta \leq \pi,$$

or

$$-\frac{\pi}{2} \leq \delta \leq \frac{\pi}{2}.$$

Usually the range of variation is chosen in accordance with the second relationship.

Although the dependence of phase on energy is determined by the properties of the interaction in an extremely complicated way, it is possible to obtain certain general results at low energies. In particular, in systems in which there are no Coulomb forces (for example, a system composed of a neutron and a proton) and in which the forces act only over small distances, the phase shift depends on the energy of the system E (or on the wave vector $\mathbf{k}$) in accordance with the formula

$$\delta_L \sim k^{2L+1} \sim \frac{1}{\lambda^{2L+1}} \sim E^{L+1/2}.$$

According to this formula the phase shifts fall off with energy and the rate of reduction is greater at higher values of L. The S-phase falls off slowest of all (i. e., the phase δ_0) and at very low energies, is the only one which remains; thus this quantity characterizes the low-energy scattering completely. Physically, this result means that at low energies the scattering becomes almost spherically symmetric and the cross section becomes independent of energy. The latter result follows from the fact that the cross section is proportional to the product of λ^2 and $\sin^2 \delta_0$, or λ^2 and δ_0^2, since δ_0 is small. However, δ_0 itself is proportional to $1/\lambda$, so that the product $\lambda^2\delta_0^2$ is independent of energy.

The general considerations which have been presented above give no information as to the sign of the phase shift at low energies. This sign depends on whether the interaction force is one of attraction or repulsion. If the particles attract each other, $\delta > 0$, and if they repel, $\delta < 0$. However, these simple considerations relating the sign of the phase to the type of interaction are valid only when the system has no bound states. In a potential well in which there are no bound states, i. e., a comparatively "shallow" well, the phase shift is positive. If the depth of the well is increased the phase shift increases, tending toward $\pi/2$ as the depth of the well approaches the value at which a level appears. Since we are limiting the range of phase variation to $-(\pi/2) < \delta < +(\pi/2)$, the phase $\pi/2$ is identical with the phase $-(\pi/2)$; with further increases in the depth of the well the phase shift increases from $-(\pi/2)$ to 0, remaining negative.

In determining phases from experimental data it should be emphasized that it is more difficult to determine the resultant sign of the phase than its magnitude. If the signs of all the phases in the scattering cross section are changed the expression itself is not changed; in this transformation the scattering amplitude becomes its complex conjugate and the square of the modulus remains the same. Thus, the sign of the phase shift cannot be found directly from scattering. To determine signs the scattering phase shifts must be related to a potential for which phase shifts can be found theoretically. Such a potential is the Coulomb potential. If we analyze a system in which there are Coulomb forces as well as the forces which we wish to investigate (nuclear forces in the present case), because of interference effects the scattering cross section can be related to the sign of the phase shifts of the nuclear forces since the phase shifts associated with the Coulomb scattering can be determined theoretically both as to magnitude and sign. In this way one determines the phase shifts in proton-proton scattering.

However, in the case of the scattering of neutrons by protons the sign of the phase shift cannot be determined in this way; the phase shift can be found only by using the fact that there is a bound state in a system comprising a proton and a neutron—the deuteron. We shall return to this problem later.

We now consider the possible forms of the interaction between nucleons when spin is taken into account. At the outset we shall limit ourselves to the case of nucleons which move with velocities small compared with the velocity of light since it is these velocities with which we are concerned in the majority of phenomena in nuclear physics. For such velocities it is reasonable to assume that the interaction potential is independent of velocity. Under this assumption we can limit the possible forms of the potential. In the absence of spin one expects that the interaction will depend only on the distance between the particles. However, in nucleons the interaction can depend on and, as has been shown experimentally, does indeed depend on the relative orientation of the spins.

We now determine the possible forms which the potential can assume. We know that the interaction must be determined by the distance between the particles r and the spin vectors s_1 and s_2. It is apparent that the potential expression will contain one term which depends only on the coordinate $V_1(r)$ and other terms which depend on the spins. Owing to the special quantum-mechanical properties of the spin vector s, for spin $1/2$ the components can enter only to the first power; any quadratic combination of the components of this vector can be reduced to one component or to a constant through the use of the following relations:

$$s_x{}^2 = s_y{}^2 = s_z{}^2 = 1/4, \qquad s_x s_y = 1/2\, i s_z, \qquad s_y s_z = 1/2\, i s_x, s_z s_x = 1/2\, i s_y.$$

Hence, only one scalar quantity $s_1 \cdot s_2$ can be formed from two spin vectors and the potential energy will contain a term $V_2(s_1 \cdot s_2)$, where V_2 is some new function of the coordinates. The remaining terms can contain the projection of the spin along the direction of the radius vector $s_1 \cdot n$ or $s_2 \cdot n$ where n is the unit vector (r/r). However, these products cannot enter individually because the spin vector s_1 is a pseudovector (or, as it is sometimes called, an axial vector). This means that in making a transformation from a right-hand system of coordinates to a left-hand system this vector reverses direction (as does any angular momentum vector). Consequently, the product $s_1 \cdot n$ is not a scalar; it is a pseudo-scalar and its sign depends on the choice of coordinate system.

Thus the potential energy, which is a scalar by definition, can contain only the product $(s_1 \cdot n)(s_2 \cdot n)$ since this is a real scalar quantity. It can be shown that it is impossible to form any other scalar expression which is independent of velocity; thus, we have the following expression for the interaction energy:

$$V(r) = V_1(r) + V_2(r)(s_1 \cdot s_2) + V_3(r)(s_1 \cdot n)(s_2 \cdot n).$$

This expression includes all possible types of interactions for two protons. However, it does not include all possible types of interactions which may exist between protons and neutrons.

It has been shown experimentally that one other process, which is unknown in classical mechanics, takes place in scattering. The particles can exchange charge during the interaction. The existence of exchange forces is related to the high degree of similarity between the proton and neutron—it is believed that when these particles are in close proximity a light charged particle is transferred from the proton to the neutron (or from the neutron to the proton), thereby changing the charge states. The existence of these charged particles is now unquestioned: these are the so-called π-mesons, about which more will be said later.

In describing the exchange interaction we proceed by analogy with the preceding potential, introducing three additional scalar functions: $V_4(r)$, $V_5(r)$ and $V_6(r)$:

$$V_{\text{exch}}(r) = [V_4(r) + V_5(r)(s_1 \cdot s_2) + V_6(r)(s_1 \cdot n)(s_2 \cdot n)]_P.$$

Here P is an operator which exchanges the positions of the neutron and proton. Thus, a description of the interaction in a proton-neutron system requires a knowledge of six functions; this is the basic problem of nuclear physics and it is far from solved. If the particles involved in the scattering process are spinless and do not exchange charges, the problem reduces to the determination of a single function; spin and charge, however, make the solution much more complicated.

It has been found experimentally that all six terms are of the same order of magnitude; thus we have no justification for neglecting any of these interactions.

The expressions for the potential can be written in a somewhat different form in which the spin of the system appears explicitly. The two particles, the proton and neutron, may exist in a state in which the total spin is 0 or in a state in which the total spin is 1. The first state is called the singlet state (multiplicity equal to 1), the second state is the triplet state (multiplicity equal to 3). We may recall that in general the multiplicity of a system is given by the number $2S + 1$, where S is the spin (the multiplicity is equal to the number of possible orientations of the spin vector S). In states with $S = 0$, it is obvious that the spin cannot play a role, and in these states the potential for each of the interactions (simple and exchange) can depend only on the distance; we designate this case by $U_1(r) + U_{1 \text{ exch}}(r)_P$. In the state with $S = 1$ the spins are parallel and the potential can be divided into two parts: one is independent of the direction of the total spin in space and is written in the form $U_2(r) + U_{2\,\text{exch}}(r)\text{P}$, while the second depends on this direction. The dependence on direction in a system of two particles implies a preferred direction—that of the vector which connects the two particles; hence, the directional dependence of the potential is given by the pseudoscalar $\mathbf{S} \cdot \mathbf{n}$.

Because of the pseudoscalar nature of the product $\mathbf{S} \cdot \mathbf{n}$ this quantity can enter the potential only in quadratic form $(\mathbf{S} \cdot \mathbf{n})^2$. Higher powers are excluded by the quantum-mechanical properties of the vector $\mathbf{S}$ for $S = 1$.

It is customary to write the potential in a more convenient form. In particular, the term which depends on the spin direction is written so that its average over direction becomes zero. Averaging the expression $(\mathbf{S} \cdot \mathbf{n})^2 = \Sigma S_i S_k n_i n_k$ over all directions of the vector $\mathbf{n}$ and using the relations

$$\overline{n_i n_k} = \tfrac{1}{3}\delta_{ik}$$

(these follow directly if it is recalled that the n_i are the direction cosines of the lines which connect the particles), we obtain:

$$\overline{\sum_{i,\,k} S_i S_k n_i n_k} = \tfrac{1}{3} \sum_{i,\,k} S_i S_k \delta_{ik} = \tfrac{1}{3} \sum_i S_i^2 = \tfrac{1}{3} S(S+1) = \tfrac{2}{3}.$$

In the last expression we have used the properties of the square of the quantum-mechanical angular momentum vector for $S = 1$.

Using the expression which has been derived we write the factor which determines the directional dependence of the potential in the form

$$[U_3(r) + U_{3\,\text{exch}}(r)\text{P}][(\mathbf{S} \cdot \mathbf{n})^2 - \tfrac{2}{3}].$$

Thus, in the singlet state the potential is

$$U_{\text{sing}} = U_1(r) + U_{1\,\text{exch}}(r)\text{P},$$

while in the triplet state

$$U_{\text{trip}} = [U_2(r) + U_{2\,\text{exch}}(r)\text{P}] + [U_3(r) + U_{3\,\text{exch}}(r)\text{P}][(\mathbf{S} \cdot \mathbf{n})^2 - \tfrac{2}{3}].$$

It is clear that the number of functions required to describe the interaction (namely, six) remains the same, regardless of the representation. There are no general relations which can be used to reduce this number. It is impossible to establish any general relation between the interactions in the singlet state and the triplet state—the connection between the functions $U_1(r)$ and $U_2(r)$ arises as a result of the properties of the actual nuclear system, consisting of the proton and neutron in the present case. Experiments indicate that the last term—the potential due to tensor forces[*]—is as important as the other terms.

One of the first experimental findings which indicated the role of the tensor forces was the existence of the quadrupole moment in the deuteron. The spin of the deuteron is unity—this is to be expected as the result of adding the two parallel spins of the proton and neutron, each equal to $\tfrac{1}{2}$. If there were no

---

[*] The designation "tensor" is due to the fact that the spin enters the potential in the form of a tensor with components $S_i S_k$.

tensor forces, the spin of the deuteron would be independent of its axis—the line which connects the proton and neutron. This would imply that the charge distribution in the deuteron is independent of the spin and the spin would not define a preferred direction; it follows that the charge distribution would be, on the average, spherically symmetric. It is well known that a spherically symmetric charge distribution gives rise to a field which is identical with the field of a point charge, i. e., such a system has no electric moments; in particular, it has no quadrupole moment. For this reason the existence of a quadrupole moment in the deuteron is to be interpreted as an indication of an asymmetry in the charge distribution.

In terms of its electrical properties the deuteron can be considered a charged ellipsoid of revolution with axis of rotation parallel to the spin; the ellipsoid is elongated rather than flattened since the quadrupole moment of the deuteron is positive.

The deuteron quadrupole moment is actually quite small. However, this situation does not arise because the interaction forces are weak but rather because the deuteron dimensions are large. Because of the small binding energy the proton and neutron are, on the average, relatively far apart (long range of nuclear forces). It can be shown by calculation that the tensor forces must be quite large to produce even a small quadrupole moment in a system with large distances between particles.

Thus, it turns out to be difficult to describe even the simplest nuclear system—a system of two particles. For this reason the study of nuclear forces is being carried out at the present time with considerably less information than is available in the case of atomic systems. In the latter the Coulomb field gives rise to comparatively simple features which are well known.

LECTURE TWO

Nuclear Forces

(SCATTERING OF NUCLEONS BY NUCLEONS)

As we have already indicated, experimental data on the proton-neutron system are acquired by studying the scattering of neutrons by protons and by investigations of the properties of the deuteron. An analysis of these data, the goal of which is the determination of phase shifts, requires an expansion of the scattered waves in terms of spherical functions. Because the proton and neutron have spins this procedure is more complicated than in the case of spinless particles, which have been considered in the last lecture. First we must introduce a system for classifying the possible states of the system. In the case of spinless particles the state is determined by the azimuthal quantum number L. If the particles have spin the state of the system is also a function of the spin quantum number S. If the spins and coordinates are independent, the orbital angular momentum and the spin angular momentum are conserved individually. Any coupling between the spin and orbital motion means that the vectors $\mathbf{L}$ and $\mathbf{S}$ are not conserved individually. Instead the total angular momentum $\mathbf{J}$ is conserved;* this quantity is the sum of the orbital and spin angular momenta.

In nuclear systems, as in all other quantum-mechanical systems, in addition to the momentum $\mathbf{J}$ there is one other quantity which is conserved: this is the *parity*, a quantity which has no classical analog. The parity determines the properties of the wave function of the system with respect to reflection through the origin of coordinates. The parity is positive if the wave function of the system does not change sign under such a reflection; in this case the system is said to be *even*. The parity is negative if the wave function changes sign under such a reflection (an *odd* system). The division of systems in terms of parity is similar to the division of vector quantities into real and pseudo quantities which was discussed earlier.

The states of a system composed of two particles are designated as in atomic spectroscopy. Capital letters are used to designate the orbital moment: the azimuthal quantum numbers $L = 0, 1, 2, 3, 4, 5$, etc., are denoted by the letters $S, P, D, F, G, \ldots$. The multiplicity index $2S + 1$ is given above and to the left while the value of $\mathbf{J}$ is given below and to the right. Thus, $^3D_3$ designates a triplet state with $L = 2$ and a total moment of 3.

---

* In what follows the quantum number designating the total angular momentum will be called J.

First we shall write out the singlet states. Since the spin in these states is zero, $\mathbf{J}$ coincides with $\mathbf{L}$ and the classification is the same as for particles without spin: $^1S_0$, $^1P_1$, $^1D_2$, $^1F_3$, $^1G_4$, etc.

In the triplet states the classification scheme is somewhat more complicated. When $L = 0$, we have only one state since the spin, equal to 1, added to $L = 0$ yields unity. This is the $^3S_1$ state. In the state with $L = 1$, we have two vectors—the orbital moment and the spin, each of which is unity. Adding these, we obtain three possible values for the total moment, 0, 1, and 2. Thus, there are three states (corresponding to the multiplicity of three): $^3P_0$, $^3P_1$, and $^3P_2$. Likewise, when $L = 2$ we obtain the triplet $^3D_1$, $^3D_2$, and $^3D_3$. Continuing this process we arrive at the triplet system:

$$^3S_1, \begin{Bmatrix} ^3P_0 \\ ^3P_1 \\ ^3P_2 \end{Bmatrix}, \begin{Bmatrix} ^3D_1 \\ ^3D_2 \\ ^3D_3 \end{Bmatrix}, \begin{Bmatrix} ^3F_2 \\ ^3F_3 \\ ^3F_4 \end{Bmatrix}, \begin{Bmatrix} ^3G_3 \\ ^3G_4 \\ ^3G_5 \end{Bmatrix} \text{ etc.}$$

The $^3S_1$ state is called a triplet state to maintain the convention although there is only one state when $L = 0$.

Having established the possible values for the angular momentum (the quantum number L) and the spin of the neutron-proton system, we can now carry out the classification of the states of this system. A good quantum number for the system is J; this quantity characterizes the total momentum and the parity of the system. In a system of two particles the parity is determined uniquely by the azimuthal quantum number L. Since the wave function of the system with azimuthal quantum number L contains the Lth Legendre polynomial $P_L (\cos \vartheta)$ and since the Legendre polynomial is multiplied by the factor $(-1)^L$ under transformation from a right-handed coordinate system to a left-handed system,* the parity of the system is determined by L and is $(-1)^L$. Thus we see the origin of the name for this quantum characteristic.

Thus, the S, D, G states are even states while the P, F states are odd states. If we now consider the possible values of J and L we find that in the triplet states the value of J and the parity, in the general case, do not uniquely determine the value of L. L is determined uniquely only in triplet states such as the $^3D_2$ state, in which J coincides with L, and in all singlet states. In triplet states in which $J = L \pm 1$ there are always two states with the same value of J and the same parity but with different L (aside from the value $J = 0$ (odd) which gives rise to only one state, the $^3P_0$ state). These two states "combine" with each other. This means that the actual state of the system is a superposition of two states with different values of L (obviously these must differ by 2 units). We may say that the system spends only part of the time in each state with different L. Thus, for example, the deuteron is a system with $J = 1$ (even) and spends approximately 96% of the time in the $^3S_1$ state and 4% of the time in the $^3D_1$ state.

---

* In this transformation ϑ goes over to $\pi - \vartheta$ and $\cos \vartheta$ to $- \cos \vartheta$. In other words, a Legendre polynomial with even L is an even function of its argument while a polynomial with odd L is an odd function. This is the basis for the statement which follows.

It is easy to enumerate the possible states for an n-p system. These states are designated by a number (the value of J) and by a superscript $+$ or $-$, denoting the parity of the system. Thus, in this classification system the deuteron state is designated by (1^+). If the state is actually a superposition of two states with different L we designate it symbolically by the sum of these states: for example, $^3S_1 + ^3D_1$ for the deuteron.

The possible states of the n-p system are shown in Table 1.

TABLE 1

Singlet	Triplet
$(0^+)^1S_0$	$(0^-)^3P_0$
$(1^-)^1P_1$	$(1^+)^3S_1 + ^3D_1$
$(2^+)^1D_2$	$(1^-)^3P_1$
$(3^-)^1F_3$	$(2^+)^3D_2$
etc.	$(2^-)^3P_2 + ^3F_2$
	$(3^+)^3D_3 + ^3G_3$
	etc.

A division has been made between singlet and triplet states. However, it does not necessarily follow from the above that singlet and triplet states cannot combine (for example, $^1D_2$ and $^3D_2$ states). However, it can be shown that if the energy of the interaction depends only on the coordinates and spin (i. e., if the interaction is velocity independent) these states cannot in fact combine.

One of the important features of the n-p system is the existence of a bound state, the deuteron. The deuteron has a binding energy of 2.23 Mev. As has already been mentioned, this quantity does not reflect the strength of the interaction between the neutron and the proton. The actual "potential well" is considerably deeper than 2.23 Mev. If we describe the interaction in terms of the potential-well model the depth of the well must be greater than 30 Mev. The relatively small binding energy indicates merely that the depth of the "well" is not much greater than that required for the existence of the first level. If the binding energy were strictly zero, the following relation between the depth of the "well" and the radius would be satisfied:

$$Vr_0^2 \sim \frac{\pi^2}{8} \frac{\hbar^2}{\mu} \approx 10^{-24} \text{ Mev} \cdot \text{cm}^2.$$

For the radius $r_0 = 2 \cdot 10^{-13}$ cm (which must be taken to agree with experiment), $V = 25$ Mev; actually, V is one and a half times larger. These numbers, among other things, illustrate the extent to which the well must be "deepened" in order to increase the binding energy from 0 to 2.23 Mev.

How should the deuteron state be classified? Experiment indicates that the spin of the deuteron, i. e., the quantum number J, is 1. The table given above contains three states with this value of J. These are the two states $(1^-)^1P_1$ and $(1^-)^3P_1$ and the state $(1^+)^3S_1 + ^3D_1$. How is the parity of the deuteron to be determined? The most reliable information is obtained from experiments on the scattering of neutrons by protons. This effect has now been investigated very carefully over a wide energy region, ranging from hundredths of an electron volt to hundreds of millions of electron volts.

It has been shown experimentally that this scattering is spherically symmetric (in the center-of-mass system) for neutron energies ranging from zero to approximately 20 Mev. This region is undoubtedly to be associated with S-scattering. Using the magnitude of the scattering phase shift at low energies and an analysis of the experimental data we can determine whether or not the system has a shallow S-level and can calculate the energy of this level if it exists. Such analyses of the scattering of neutrons by protons have shown that the system has a level corresponding to the $^3S_1$ state and that this level has an energy equal to the binding energy of the deuteron. Actually, there is a small admixture of the $^3D_1$ state which does not affect scattering at low energies.

Scattering experiments also determine the phase shift associated with the $^1S_0$ state; this phase shift turns out to be large at very small energies because the scattering at these energies is of resonance nature. This means that the interaction is very similar to that in the system in which there is a level at $E = 0$. The answer to the question of whether there actually is a level in the system depends on the sign of the phase shift; if a level does exist this phase shift should be negative. However, it is impossible to determine the sign of the phase shift from the scattering of neutrons on free protons—the cross section does not change if the sign of the phase shift is reversed. To determine the signs it is necessary to investigate effects in which interference takes place between scattering in the $^1S_0$ and $^3S_1$ states. Such an effect is the scattering of slow neutrons on hydrogen molecules in which the spins of both nuclei are parallel (ortho-hydrogen) or anti-parallel(para-hydrogen). It has been established from these experiments that the sign of the phase shift is different in the $^3S_1$ and $^1S_0$ states. Since there is a level corresponding to the $^3S_1$ state (deuteron), it follows that the phase shift is positive in the $^1S_0$ state and there is no level.

Although there is no level associated with the $^1S_0$ state, in analyzing phase shifts it is convenient to introduce the idea of a "virtual" level.

Let V_0 be the depth of the well for which a level exists (i. e., the depth for which the level energy is zero). Then, a well of depth $V_0 + \delta V$, where δV is small, will have a level, while a well of depth $V_0 - \delta V$ will not. The scattering phase shift at zero level-energy for a potential V_0 is by definition $\pi/2$. If the well is now made somewhat deeper (potential $V_0 + \delta V$) the phase becomes somewhat larger. According to the convention which has been adopted the phase shift must lie within the interval $[-(\pi/2), \pi/2]$. Hence, an increase in phase means that it becomes **negative**, assuming a value close to $-(\pi/2)$. On the other hand, the potential $V_0 - \delta V$ obviously leads to a **positive** phase with a value close to $\pi/2$.

A more exact analysis shows that in both of these wells the scattering at low energies is of the same magnitude, but of opposite phase. We can imagine that the real level in the $V_0 + \delta V$ well corresponds to a "virtual" level of the same magnitude in the $V_0 - \delta V$ well. This terminology is especially convenient since the scattering cross section at low energies is independent of the sign of the phase shift; that is to say, it does not depend on whether the level is real or virtual (scattering from a $V_0 + \delta V$ well and a $V_0 - \delta V$ well is the same).

It turns out that the depth of the "virtual" level for the $^1S_0$ state is approximately 70 kev.

Thus, at low energies a scattering analysis makes it possible to find a parameter which characterizes the interaction from the measured cross section. It also turns out that an investigation of the energy dependence of the scattering allows us to find one more parameter. Hence, two parameters can be determined from experiment. These parameters acquire more significance when used to describe the interaction in terms of some concrete model. For example, if one chooses a rectangular well, the depth and width can be found experimentally.

Scattering theory indicates that it is impossible to determine any more than two constants from the scattering of low-energy neutrons. Hence, the true shape of the potential cannot be found in experiments with low-energy neutrons even in the simple case of the singlet state in which the interaction, as has already been pointed out, is described by a single function.* The experimental data can be fitted to almost any potential as long as the interaction vanishes at large distances. Different forms of the two-parameter potential are used in different work. Typical examples are a rectangular well (with parameters radius r and depth V, as described above) and a potential of the form

$$V_0 \frac{e^{-\alpha r}}{r}$$

with the parameters being V_0 and α. The actual shape of the potential has still not been determined because even experiments with high-energy neutrons provide no data for solving this problem. Consideration of the concrete form of the potential can, at present, serve only to give a picture of the interaction. For instance, it can be shown that scattering in the $^1S_0$ state can be described by a rectangular well with a radius of $2.8 \cdot 10^{-13}$ cm and a depth somewhat greater than 10 Mev.

The situation is much more complicated in the triplet state $^3S_1$. In the triplet states, as has been pointed out above, the tensor forces play an important role and the $^3S_1$ state is mixed with the $^3D_1$ state. If, for the sake of argument, we neglect the tensor forces, this state can be described by scattering in a rectangular well. In this case the triplet-state "well" is much narrower than the singlet well and the depth is approximately three times greater. These numbers give some idea of the effect due to the relative orientation of the spins of the colliding particles.

Slow-neutron experiments give no information which can be used to estimate the magnitude of the tensor forces. Only one quantity is available for making such an estimate—the quadrupole moment of the deuteron, and this does not provide sufficient information for a quantitative description. It can only be stated that the tensor forces are of approximately the same order of magnitude as the ordinary forces.

---

* In the $^1S_0$ state this function is the sum of the potentials $V_1 + V_{1\ \text{exch}}$.

We now turn to a second system: the p-p system. It differs from the n-p system primarily in that there is also an electric interaction between protons. The presence of the nuclear interaction changes the distribution for proton-proton scattering from that which is calculated on the basis of a Coulomb law (*Rutherford scattering*).

Identity considerations must be taken into account when two protons interact. This is in accordance with the requirement imposed on such a system by the Pauli principle. According to the Pauli principle the wave function for a system of two protons must be anti-symmetric with respect to interchange of coordinates and spins for both particles (it must change sign in such an interchange). In triplet states of the system the wave function does not change sign upon interchange of particle spins; in the singlet states the wave function does change sign in such an interchange. To make the total wave function anti-symmetric it must change sign upon interchange of particle coordinates in triplet states (the states must be odd states) and must not change sign for interchange of coordinates in singlet states (the singlet states must be even states). It follows from these considerations that the p-p system has available only half the states which are allowed for the n-p system (in which there is no limitation on the parity of the state and the symmetry of the wave function).

The allowed states for a p-p system are shown in Table 2.

TABLE 2

Singlet	Triplet
$(0^+)^1S_0$	$(0^-)^3P_0$
$(2^+)^1D_2$	$(1^-)^3P_1$
$(4^+)^1G_4$	$(2^-)^3P_2 + {}^3F_2$
etc.	$(3^-)^3F_3$
	etc.

Proton-scattering experiments at energies of several millions of electron volts yield important results concerning the interaction in the $^1S_0$ state. A scattering analysis allows us to reduce the experimental data to the determination of two constants. An important experimental fact, discovered in proton-proton scattering experiments, is that the proton-proton and proton-neutron interactions are virtually identical.

If we write the interaction potential for two protons in the $^1S_0$ state in the form

$$\frac{e^2}{r} + V(r),$$

i. e., as a sum of the Coulomb potential and the nuclear potential, the parameters for the potential $V(r)$ virtually coincide with the parameters for the potential $V_1 + U_{1\ \text{exch}}P$ in the $^1S_0$ state of the n-p system. Hence, it follows that there is no stable level in the p-p system (just as in the $^1S_0$ state of the n-p system). In other words, the nucleus He^2 cannot exist in nature.

The similarity of the interaction in p-p and n-p systems is of fundamental significance. This similarity indicates the deep-lying symmetry between these particles. This symmetry goes far beyond mere charge symmetry; it appears in the identical spectra of mirror nuclei and gives rise to the identical forces between protons and between neutrons. All the presently available data on nucleon behavior at all energies indicate that p-p, p-n, and n-n systems are very similar, not only in scattering in the $^1S_0$ state but in all effects in which the Coulomb force is relatively unimportant. This property is called *isotopic* (or charge) *invariance*.

It should be emphasized that isotopic invariance by no means implies complete equivalence in the interactions for the three different particle pairs; for one thing the n-p system has states which cannot exist in identical-particle systems. It is clear that the interactions in these states are not directly related to the properties of the p-p or n-n systems. In speaking of isotopic invariance we are discussing the interaction in identical states of different systems.

In considering isotopic invariance it is convenient to introduce a formalism which will be found very useful in obtaining information as to the nature of different nuclear systems. We present the basic features of this formalism, which is purely an abstract description.

The starting point of this scheme lies in considering the neutron and proton as two different "charge states" of the same particle. These two charge states of the nucleon are described by means of a new concept—*isotopic spin*.* In this scheme the particles are characterized by a new vector—the isotopic spin vector τ; its absolute value is $1/2$. As for ordinary spin $1/2$ its projection along a preferred axis in isotopic spin space—the ζ axis—can have only two values (the other two axes are designated by ξ and η). The value $+1/2$ indicates that the particle is a proton, and the value $-1/2$ indicates that the particle is a neutron. The transition from neutron to proton is denoted by a change in the value of the projection from $-1/2$ to $+1/2$. By analogy with ordinary space this operation can be described as a rotation through 180° about the ξ axis in isotopic spin space. (In ordinary space this would be a rotation about the x or y axes.) If the system consists of several nucleons the isotopic spins are added in accordance with the rules for quantum-mechanical addition of vectors. In such a system the projection of the total isotopic spin on the ζ axis is equal to the neutron excess† $N - Z$ multiplied by $-1/2$, since each proton contributes $+1/2$ to the projected value and each neutron contributes $-1/2$.

The concept of an isotopic spin space is purely formalistic and abstract. Hence, it is meaningless to discuss the relation between this space and ordinary space; similarly, it is quite pointless to try to extend the interpretation given above.

The convenience of isotopic spin space lies in the fact that the isotopic invariance features can be formulated compactly. If we neglect Coulomb forces, a system consisting of two protons differs from a system consisting of one proton

---

* Isotopic spin = isobaric spin in more recent American usage.—Translator.

† Z is the atomic number and N is the number of neutrons.

and one neutron only by a rotation in isotopic space. Analogous states of these two systems (that is, states with the same quantum numbers) can be considered as states corresponding to different projections of the isotopic spin of the two-nucleon system along the ζ axis in isotopic space. In these terms isotopic invariance is described in terms of the invariance of a system with respect to rotation in isotopic space. We may note that charge symmetry—invariance of the properties of the system under replacement of all neutrons in the system by protons, and vice versa—appears as a particular case of isotopic invariance; in this formalism we say the system is invariant under a rotation of 180° about an axis in the $\xi\eta$ plane in isotopic space.

Obviously, isotopic invariance does not extend to the Coulomb interaction. This can also be shown in a formal way: the charge is given by the projection of the isotopic spin; consequently the Coulomb interaction depends on the projection of the isotopic spin, a quantity which is obviously not invariant against rotation of the system of coordinates.

The introduction of isotopic spin makes it possible to extend the classification system for the states of two-nucleon systems through the introduction of a new quantum number, the isotopic spin. Since each of the nucleons has an isotopic spin equal to one-half, according to the rules for adding quantum-mechanical vectors, the vector sum of these must be either unity or zero. Hence, all states of the system can be divided into two groups: one is associated with isotopic spin $T = 1$, the other with $T = 0$.

An isotopic spin $T = 1$ can have three values for its projection. In accordance with rules given above, the projection $T_\zeta = -1$ is to be associated with a charge 0 for the system, the projection $T_\zeta = 0$ with a charge $+1$ (in units of e, the charge of the electron) and the projection $T_\zeta = 1$ with charge $+2$. It is apparent from these considerations that a system with $T = 1$ can be realized by two neutrons, two protons, or a neutron and a proton; in the latter case we can associate with states $T = 1$ only those states of the p-n system which can also exist in a system of two identical particles, i. e., states which satisfy the Pauli principle.

The state with $T = 0$ can have only one projection, $T_\zeta = 0$; this describes all states of the n-p system which do not satisfy the Pauli principle. It is apparent that these states are described by a symmetric wave function (these states are even triplets and odd singlets).

Finally, we arrive at the following classification of states:

$$\text{for } T = 1 \qquad {}^1S_0,\ {}^3P_{0,\ 1,\ 2},\ \text{etc.};$$
$$\text{for } T = 0 \qquad {}^3S_0,\ {}^1P_1,\ \text{etc.}$$

It is interesting to note that in this scheme the triplets and singlets (for a given L) always refer to different values of T, and hence cannot combine with each other.

Using the idea of isotopic spin, it is now possible to formulate exactly the properties of isotopic invariance of nuclear forces as applied to a system of two nucleons: the isotopic invariance now expresses the fact that the interaction is the same for any two nucleons in states with $T = 1$.

Before turning to a description of the experimental results with high-energy nucleons, something should be said about nucleon interactions at low energies in states with $L > 0$.

We have already indicated that it is only at low energies that the interaction is basically described in terms of the S-state. Theoretically, it is impossible to predict at what energy other states start to appear. It has been shown experimentally that even at 20 Mev the role of these other states is still small and a determination of the residual phase shift remains difficult. At 40 Mev other states play an important role in the p-n system. It is unfortunate that it is precisely this energy region (20 to 40 Mev) which is almost completely uninvestigated. At an energy close to 20 Mev, the phase shift in the P-state is estimated as approximately 1° in the p-p system; because of the small value of this phase shift it is impossible to determine which of the three P-states is responsible.

Nothing more can be said with regard to this state. Experiments, which we will discuss later on, indicate that the interactions in all three P-states ($^3P_0$, $^3P_1$, $^3P_2$) are considerably different, and hence any statement as to the average interaction in the three P-states would have very little validity. Still less can be said concerning the parameters for this interaction. The interaction is characterized here by the product of the well depth and the fifth power of the radius (in the S-state it would be the square of the radius); hence, the well depth is very sensitive to the choice of radius.

Nuclear Forces

(SCATTERING OF NUCLEONS AT HIGH ENERGIES)

We have seen that experiments on nucleon-nucleon scattering at low energies yield information only on the interaction in $^3S_1$ and $^1S_0$ states and are not very useful in determining the interaction potentials between nucleons. The reason lies in the fact that at energies of the order of 10 Mev the nucleon wavelength is of the order of the interaction radius and the scattering is insensitive to the detailed dependence of the interaction on distance. In order to "probe" the potential well the wavelength must be considerably smaller than the dimensions of the well. Hence, more information is obtained from the experiments with high-energy nucleons (several hundreds of millions of electron volts) which have been carried out in the last several years in various laboratories.

These experiments are all the more interesting because quantum mechanics shows that it is possible to obtain direct information on the potential which acts between the particles from the scattering data at these energies without recourse to a phase-shift analysis. This possibility is based on the assumption that the interaction energy is small compared with the particle energy; this assumption is certainly valid at high energies. However, we may not be justified in assuming that the interaction is "weak"; the picture at high energies may not be the same as that which would be expected on the basis of ordinary perturbation theory.

Nevertheless, at the outset we base our analysis of scattering at high energies on perturbation theory since this approach turns out to be useful in the later analysis.

When the kinetic energy of the colliding particles is much greater than the interaction energy, the scattering amplitude in the center-of-mass system (c.m.) is given by the following expression, which is derived in quantum mechanics (the *Born approximation*):

$$f(\vartheta) = \frac{\mu}{2\pi\hbar^2} \int V(r)e^{i(\mathbf{k}-\mathbf{k}')\mathbf{r}}d^3r.$$

In this expression μ is the effective mass, $V(r)$ is the interaction potential, $\mathbf{k}$ and $\mathbf{k}'$ are the wave vectors before and after scattering, ϑ is the angle between $\mathbf{k}$ and $\mathbf{k}'$ (scattering angle) and d^3r is a volume element.

23

We may recall that the scattering cross section is given in terms of the scattering amplitude by the expression

$$d\sigma(\vartheta) = |f(\vartheta)|^2 do$$

and that ϑ and $\mathbf{q} = \mathbf{k} - \mathbf{k}'$ are related by the expression (for elastic scattering)

$$\mathbf{q}^2 = (\mathbf{k} - \mathbf{k}')^2 = 4k^2 \sin^2 \frac{\vartheta}{2}.$$

It is easy to show that the scattering amplitude (to within a multiplicative constant) is simply the Fourier amplitude of the interaction potential. Thus, if the scattering theory formulas apply, having determined the scattering amplitude, we can find the potential through the properties of the Fourier integral.

These formulas describe the angular distribution of the scattered particles. For purposes of illustration we choose a potential which differs from zero only in a region of dimensions a.

In the scattering amplitude formula the factor $e^{i\mathbf{q} \cdot \mathbf{r}}$ under the integral sign is a periodic function with period (in the direction of the vector $\mathbf{q}$) $2\pi/q$. If this period is small compared with the dimensions of the well, this oscillating factor changes sign many times within the boundaries of the well. Thus the integral in the expression for $f(\vartheta)$ will be small since $V(r)$ varies at a much slower rate than $e^{i\mathbf{q} \cdot \mathbf{r}}$ and the different parts of the integral will approximately cancel. The larger the value of q, the more effective is this cancellation effect; hence when $q \gg 2\pi/a$ the value of the integral may be taken as zero. This integral will be different from zero only for

$$q \lesssim \frac{2\pi}{a}.$$

On the other hand, at small ϑ the magnitude of the vector $\mathbf{q}$ is related to the scattering angle by the expression

$$q \approx k\vartheta.$$

Whence it is obvious that the integral will be nonvanishing only for $\vartheta < 2\pi/ka$.

Thus, the angle ϑ, within which the particles are scattered, is inversely proportional to the wave vector, i. e., the square root of the particle energy. In this case, the amplitude corresponding to scattering in the direction $\vartheta = 0$ is independent of particle energy and is given by the expression

$$f(0) = -\frac{2\mu}{\hbar^2} \int V(r) r^2 dr.$$

It can be shown from perturbation theory that at high energies the scattering cross section is constant only within a small solid angle, $\pi\vartheta^2$, which (in the c.m. system) is inversely proportional to the energy of the particle. Under these conditions, the total scattering cross section is obviously inversely proportional to E.

The scattering picture described above is changed if the forces which act between the particles have exchange character. The nature of this change is easily understood if one considers, in the problem which has just been discussed, the angular distribution of the recoil particles rather than the scattered particles (the recoil particles are those which are at rest in the laboratory system before the collision). In the center-of-mass system the recoil particles move in a direction opposite to that of the scattered particles and thus the angular distribution of these particles is described by the same scattering cone but in the backward direction (the region $\vartheta \sim \pi$).

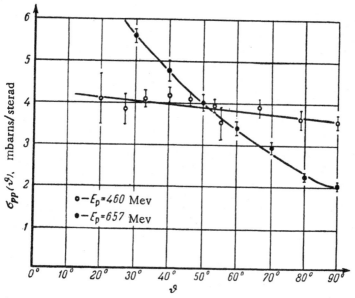

Fig. 3.   Elastic proton-proton scattering (from data reported by N. P. Bogachev, I. K. Vzorov, M. G. Meshcheryakov, B. S. Neganov and E. V. Piskarev). Proton energy 460 Mev (almost isotropic scattering) and 657 Mev (highly anisotropic scattering); ϑ is the proton scattering angle in the center-of-mass system.

Exchange scattering means that charge is exchanged between particles during the scattering process; that is, an incoming neutron is scattered as a proton, while the recoil particle proves to be a neutron.

It is clear that the neutron distribution in exchange scattering is the same as the distribution of recoil particles in ordinary scattering. In the general case the scattering diagram consists of two directional cones: these are in the forward and backward directions (in the center-of-mass system), and the relative intensities are determined by the fraction of exchange force in the interaction between the particles.

The experimental results published in the last few years are found to be completely unlike the picture which has been given. A clearcut contradiction is found in experiments on the scattering of protons by protons. These experiments have shown that over an extremely wide energy range (150 to 350 Mev) the cross section is independent, within the experimental errors (approximately 10%), of both angle and energy. It is only in the region of extremely small angles that the cross section increases and this effect is due to ordinary Coulomb scattering. The cross section per unit solid angle is approximately

$$\sigma(\vartheta) = 4 \text{ mbarns/sterad.}^*$$

It is only at higher energies (430 Mev) that a noticeable angular dependence is found. In recent experiments carried out with the Soviet synchrocyclotron the scattering cross section has been investigated at 657 Mev (Fig. 3). At these energies the cross section shows a pronounced angular dependence.

These experimental results indicate that the basic assumption of perturbation theory—that the interaction is weak at high energies of the incoming particles—is not justified. The experimental results indicate that extremely powerful forces come into play when the distance between particles is small. To carry out a more detailed analysis of these forces, we must make a phase-shift analysis of the angular distribution. At the present time, however, the data which have been accumulated are still inadequate for a complete analysis of this type. Nevertheless, it is possible to show why the cross section is independent of scattering angle over a wide angular region.

It has been shown above that there is a limitation on the magnitude of the scattering cross section. In particular, the cross section for S-scattering (that is, for $L = 0$) per unit solid angle cannot exceed

$$\frac{d\sigma_{\max}}{do} = \lambda^2$$

(this formula applies both for neutron scattering and proton scattering in S-states).

At 380 Mev, λ^2 is approximately 2.5 mbarns; this value is smaller than the observed cross section. It might be said (and this statement is encountered in the literature) that there is a clearcut inconsistency. However, such is not the case. There is one other state (in addition to the S-state) which is characterized by spherical symmetry in scattering—the ${}^3P_0$ state. The maximum cross section for this state is also λ^2; thus, the states ${}^1S_0$ and ${}^3P_0$ together can give rise to an isotropic cross section of $2\lambda^2$, which is consistent with experiment. It is obvious that the phase shifts of the other states with $L = 1$ (${}^3P_1$ and ${}^3P_2$ states) must either be small or have highly anisotropic cross sections. From these considerations we conclude that the interaction is essentially different in the ${}^3P_0$, ${}^3P_1$ and ${}^3P_2$ states; that is, the spin-orbit interaction is large in a system of two protons. This behavior in a system of two particles has already been encountered in the deuteron. In that case it is responsible for the small quadrupole moment associated with mixing of ${}^3S_0$ and ${}^3D_1$ states. While the effect in question is small in the deuteron, at higher energies it becomes predominant.

---

* 1 barn = 10^{-24} cm².

Further information on the proton-proton system can be obtained from investigations of scattering of polarized particles on nonpolarized particles and scattering from polarized targets. The problem of making a target of polarized protons is an extremely difficult one and has still not been solved. It is only in the last few years that the first experiments on the scattering of polarized particles on unpolarized targets have been reported.

When a polarized beam is scattered on a target (analyzer) there is a difference in the scattering intensity in the forward and backward directions. The difference in the forward and backward cross sections also yields information as to the polarization of the beam. At the present time the results of a number of experiments on the scattering of polarized protons by protons have been reported. Although these data are as yet not very accurate, the importance at these energies of states other than $^1S_0$ and $^3P_0$ states is apparent. At about 200 Mev the phase shift of the $^3P_2$ state reaches 7–10°, corresponding to a polarization of approximately 20%. At higher energies the polarization increases and the interpretation of the data at 400 Mev requires that we consider states with $L > 1$ in addition to states with $L = 1$. These results are still far from reliable; further refinement will require further experimental developments. Additional information on proton-proton interactions should be given by experiments in which polarized targets are used in conjunction with polarized proton beams. However, work with unpolarized targets can be useful if the experiments are improved so that observations on the polarization of the recoil proton are carried out together with observations of the polarization of the scattered protons. The results of such experiments should make it possible, for example, to distinguish between scattering in $^1S_0$ and $^3P_0$ states; such a distinction is not possible in ordinary scattering experiments (as we have seen, both of these states are characterized by isotropic scattering).

An important feature of high energy scattering (higher than 400 Mev) is inelastic scattering, i. e., scattering accompanied by the production of π-mesons. Because of the production of π-mesons the proton-proton interaction changes markedly at energies in the vicinity of 1000 Mev. The total cross section for the interaction (elastic plus the production of π-mesons) increases, reaching a value of approximately 50 mbarns at 800 Mev and remaining constant [within the limits of the errors (10%)] up to 1000 Mev, the highest energy at which these measurements have been carried out. The elastic scattering, however, remains virtually unchanged at approximately 25 mbarns (the value at the lower energies). Thus the elastic and inelastic cross sections are essentially equal and constant in this region. It can be shown theoretically (we present these calculations in the Eighth Lecture) that this behavior is characteristic of scattering of particles on an absolutely black sphere, that is to say, a sphere which absorbs all particles incident on its surface. Because of the wave nature of the proton flux this absorption is accompanied by diffraction at the edges of the sphere (analogous to the diffraction of light at the edge of a black obstacle); this effect is also observed in elastic scattering.

The absorption cross section (production of π-mesons) in this model is πR^2; this is also the value of the diffraction cross section for elastic scattering. If this

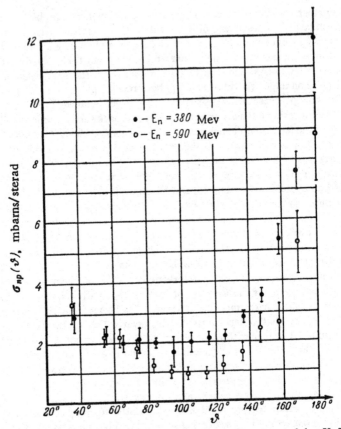

Fig. 4. Elastic neutron-proton scattering (from data reported by V. P. Dzhelepov and Yu. M. Kazarinov). Neutron energy (in the laboratory system) 380 and 590 Mev; ϑ is the neutron scattering angle in the c.m. system.

model is used to describe the scattering of protons by protons, taking the sum of the "radii" of the two protons as the collision parameter, the proton radius is found to be

$$r_p = 4.5 \cdot 10^{-14} \text{ cm}.$$

It is interesting to note that these results are in agreement with the results of experiments on the scattering of electrons by protons. Such experiments indicate that at distances smaller than approximately $7 \cdot 10^{-14}$ cm the electric field of the proton does not correspond to that of a point charge.

Elastic scattering of particles on a black sphere should also be characterized by an angular distribution typical of diffraction. Such a distribution is observed experimentally. It will be extremely interesting to examine the behavior of the cross section at higher energies.

We now turn to the scattering of neutrons by protons. This type of scattering has been studied over a wide energy range (up to 590 Mev) and yields a picture which is considerably different from that described above. The general energy dependence of the cross section may be described as follows: the cross section is a strong function of angle, increasing by several times at 180° as compared with the cross section at 90° (Fig. 4). Thus, for example, at 380 Mev the 90° cross section is 2 mbarns while the 180° value is 12 mbarns. At higher energies the 90° cross section falls off, becoming 1 mbarn at 590 Mev; however, the cross section at 180° remains almost constant (measurements reported by V. P. Dzhelepov and Yu. M. Kazarinov).

Below 90° the cross section does not increase so rapidly and the curve is asymmetric (till quite recently it had been assumed, without justification, that the scattering curve was symmetrical about 90°).

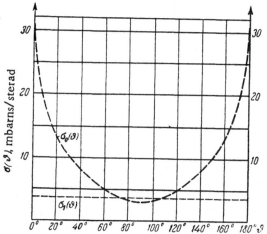

Fig. 5. Approximate dependence of the nucleon scattering cross section in states with isotopic spin $T = 0$ and $T = 1$. The nucleon energy is approximately 400 Mev; ϑ is the scattering angle in the c.m. system.

There is good reason to believe that high-energy experiments are completely in agreement with the hypothesis of isotopic invariance of nuclear forces. The isotopic spin formalism allows us to make some progress in an analysis of the results of these experiments.

As indicated above, we can assign an isotopic spin of $1/2$ to a nucleon. The proton corresponds to an isotopic spin projection of $+1/2$ while the neutron corresponds to a projection $-1/2$. A system of two nucleons can have an isotopic spin of either unity or zero. In a system consisting of two protons the isotopic spin is unity (the projection is $+1$). An n-p system may be found in a state with isotopic spin of either 1 or 0. In states with $T = 1$ the interaction should be the same as that for the interaction of two protons (neglecting the Coulomb interaction). States with $T = 0$ do not have an analog in a system of two protons and the interaction in these states applies only to two different particles. Hence, even within the framework of isotopic invariance one would not expect that the scattering of neutrons by protons would be the same as the scattering of protons by protons. Moreover the neutron cross section at a given angle should not be equal to the sum of the scattering cross sections for the states with $T = 0$ and $T = 1$ since these states interfere with each other. It might be possible, however, to find a method for identifying the interference terms, so that it would be possible to distinguish between these states. For this purpose, instead of using the neutron cross section at ϑ in the c.m. system we take the sum of the cross sections at ϑ and $\pi - \vartheta$. The physical significance of this procedure is that the neutron cross section is replaced by the particle cross section, that is to say, we consider the total number of neutrons and protons scattered at a given angle. This procedure is a natural one in the framework of isotopic invariance, and the interference terms vanish.

The quantity

$$\sigma'(\vartheta) = \sigma_{np}(\vartheta) + \sigma_{np}(\pi - \vartheta)$$

then represents the sum of the scattering cross sections for the states with $T = 0$ and $T = 1$,

$$\sigma'(\vartheta) = \sigma_0(\vartheta) + \sigma_1(\vartheta).$$

Since $\sigma_1(\vartheta)$ is nothing more than the proton-proton scattering cross section (neglecting the Coulomb scattering), knowing $\sigma_{pp}(\vartheta)$, from the experimental data we are now able to calculate $\sigma_0(\vartheta)$.

First of all we must note that the cross section $\sigma_0(\vartheta)$ calculated in this way is always positive.

In Fig. 5 we show the behavior of $\sigma_0(\vartheta)$ and $\sigma_1(\vartheta)$ at energies near 400 Mev. The fact which is most striking is that these curves are completely different. The $\sigma_0(\vartheta)$ curve is very much like the curve given by perturbation theory. Although the accuracy for this curve is small (since it represents the differences between two curves, each of which is characterized by an error of the order of 10%), it is reasonable to assume that the interaction in states with $T = 0$ is weak and can be described, to some degree at least, by perturbation theory.

To close this brief survey of nuclear forces we would like to call attention to the fact that the interactions between protons and neutrons in a nucleus may be different from the interactions of free particles which are studied in scattering experiments. In ordinary atomic systems the electronic interaction is made up of the mutual interactions of all particle pairs. For example, the potential energy associated with three electrons is given by the function

$$V(r_1, r_2, r_3) = +e^2 \left(\frac{1}{|r_1 - r_2|} + \frac{1}{|r_2 - r_3|} + \frac{1}{|r_3 - r_1|} \right),$$

i. e., the interaction of the three particles is equal to the sum of the mutual interactions of the three particle pairs. This "additivity" of the Coulomb forces is a consequence of the linear equations of electrodynamics. We have no basis for believing that nuclear forces are also additive; the interaction of two nucleons may not be independent of the presence of a third nucleon in the immediate vicinity.   It is possible that nuclear forces are not additive and that the interaction of particles in the nucleus is not simply the sum of the pair interactions. If such is the case the study of complex nuclei will lead to information on nuclear forces which cannot be obtained from investigations of two-particle systems.

Nuclear Structure

(INDEPENDENT PARTICLE MODEL)

It is clear that no simple model can represent all the properties of an extremely complicated quantum system such as the nucleus. Any model, of necessity, must have limited application. We should not be surprised if different effects require different models for their description; sometimes these models may even have mutually exclusive properties (for example, the independent particle model and the optical model).

Many properties of the nucleus can be explained by the independent particle model. This model is based on an analogy with the electronic shells of atoms. The analysis of electrons in an atom is based on the assumption that each electron moves in some average field produced by all the other electrons.

In a similar way we may use the notion that each nuclear particle moves in the average field produced by all the other nucleons in the nucleus without interacting with any of the other particles individually.

We have already directed attention to the fact that the nucleon dimensions are of the order of $4.5 \cdot 10^{-14}$ cm. The distance between nucleons in a nucleus is approximately $1.8 \cdot 10^{-13}$ cm. Thus, roughly speaking, the nucleons occupy $1/60$ of the volume of the nucleus. It is not surprising to find that nucleon properties are maintained inside the nucleus. In particular, this situation is responsible for the fact that the magnetic moments of nucleons inside nuclei are the same as for free nucleons.

If, in accordance with the above, we describe the motion of each nucleon individually in a common field, it is most natural to take this field as spherically symmetric and to describe the states of the nuclear particles in terms of the well-known quantum-mechanical relations for the motion of particles in such a field. We will discuss the validity of this assumption later; in any case, this assumption certainly holds in light nuclei.

In a field of spherical symmetry a particle has a definite angular momentum. The orbital momentum of the particle (in units of $\hbar$) is denoted by the symbol $\mathbf{l}$ (in contrast to the symbol $\mathbf{L}$, which is used to denote the orbital momentum of the system). As is well known, the corresponding quantum number l can assume integral values $0, 1, 2, \ldots$, and these states are denoted, respectively, by the symbols s, p, d, f, and so on.

In addition to its orbital momentum, a nucleon has spin. Thus, the total angular momentum of the nucleon **j** is composed of two vectors

$$\mathbf{j} = \mathbf{l} + \mathbf{s}.$$

In the analysis of the interaction of two particles, we have seen that the spins of the nucleons interact with the orbital moment. This interaction can be described in terms of tensor forces, i. e., forces whose potentials are proportional to the expression

$$(\mathbf{s}_1 \cdot \mathbf{n})(\mathbf{s}_2 \cdot \mathbf{n})$$

where **n** is a unit vector along the line connecting the particles.

It is impossible to form an expression of this type for a nucleon which moves in a central field. The field itself has no spin and thus we have at our disposal only the nucleon spin **s** and its radius vector **r** or the corresponding unit vector $\mathbf{n} = \mathbf{r}/r$. It is impossible to form a scalar from these two vectors because, as we have seen, $\mathbf{s} \cdot \mathbf{n}$ is a pseudoscalar while $(\mathbf{s} \cdot \mathbf{n})^2$ is equal to $1/4$, since the projection of **s** in any direction (in particular in the direction of **n**) is $\pm 1/2$. But this does not mean that the energy of the nucleon is completely independent of the spin direction.

The spin of a nucleon in a nucleus is related to its orbital motion; the relation, however, depends on the velocity of the nucleon **v**. Hence, we can introduce one more vector—the nucleon velocity—in the energy expression. It is now possible to form a scalar from the three vectors **s**, **n** and **v**. This scalar will be of the form

$$f(r)(\mathbf{r} \times \mathbf{v}) \cdot \mathbf{s}$$

where $f(r)$ is a scalar function of the coordinates. Since

$$\mathbf{r} \times \mathbf{v} = \frac{1}{m}\,\mathbf{l},$$

the energy expression for the spin-orbit coupling can be written in the form

$$\frac{1}{m}\,f(r)\mathbf{l} \cdot \mathbf{s}.$$

This interaction is relativistic (it vanishes when $v \to 0$) and is weak compared to the interaction of the nucleon with the average field of the nucleus.

The experimental data indicate that the spin-orbit interaction is characterized by an energy of the order of 2 Mev whereas the interaction with the nuclear field is of the order of 20–30 Mev.

In a system of several nucleons the spins of the different nucleons interact with each other as well as with the orbital moments.

It is well known that in light atoms the spin-orbit interaction for a given electron is small compared with the interaction of electrons between themselves. In a system of this type the orbital moments of the particles are added, forming the total orbital momentum of the atom

$$L = \sum_i l_i.$$

The particle spins are added the same way since the coupling due to the identity of the particles is also stronger than the spin-orbit interaction

$$S = \sum_i s_i.$$

If there is no spin-orbit interaction, in general the state of the system does not depend on the mutual orientation of the vectors L and S, that is, the vector sum

$$J = L + S.$$

The weak interaction between the spins and the orbital motion means that different values of the quantum number J correspond to different states of the system.

This coupling scheme is called the *Russell-Saunders* scheme.

There is another type of coupling scheme; this scheme corresponds to the case in which the spin-orbit interaction of each individual particle is large compared with the interaction of the spins and orbits with those of other particles. In this scheme the total angular momentum of a particle is given by

$$j_i = l_i + s_i$$

and is unique since the coupling between the vectors l and s is not disturbed by the weak interaction with other particles.

The j_i vectors of the individual particles are added to form the total angular momentum of the system

$$J = \sum_i j_i.$$

This scheme is called the *j-j coupling* scheme. It has been shown experimentally that this type of coupling obtains in certain heavy atoms. In nuclei the *j-j* coupling scheme is much closer to the actual picture than the Russell-Saunders scheme.

The notation of the *j-j* coupling scheme is now described briefly.

The state of each particle is given by its total momentum and parity. In general, the orbital moment is not known in the case of arbitrary spin. However, in the case of a particle with spin $1/2$ the orbital momentum is uniquely determined. It is easy to show that the vector j can be formed either from a state with $l = j - 1/2$ or from a state with $l = j + 1/2$. For a given value (half integral!) each state has different parity; hence the parity and j determine the quantum number l.

It is obvious that the following states can exist (with different j and l): $s_{1/2}$, $p_{1/2}$, $p_{3/2}$, $d_{3/2}$, $d_{5/2}$, and so on. The $s, d, \ldots$ states are even, while the $p, f, \ldots$ states are odd. Several different states of the same type, that is, several states with the same l and j, can exist in the nucleus. They differ in terms of the numbering sequence. There are two notation systems. One of these follows the example of the principal quantum number in the hydrogen atom* and starts the enumeration of states with $n = l + 1$. In this scheme the first s-level is denoted by $1s$, the first p-level by $2p$, the first d-level by $3d$, and so on.

In the other scheme (which we will use) the numbering of the levels starts with unity.

Because we have so little information about the average nuclear field in which the nucleons move we cannot determine theoretically the order and disposition of the levels. This information can be obtained only by experiment. Hence an analysis of the nuclear properties can only establish a number of gross features of the arrangement of the levels.

It turns out that the level energy increases with increasing orbital quantum number. This rule is a result of the fact that the centrifugal potential of the particle increases with increasing l, and this leads to a reduction in binding energy.

Further, the spin-orbit coupling in the nucleus turns out to be such that the level with $j = l + \frac{1}{2}$ (that is, l parallel to **s**) is lower (has greater binding energy) than the level with $j = l - \frac{1}{2}$. There does not seem to be any exception to this rule.

The following empirical rule pertains to the isotopic spin of nuclei.

We recall that the projection of the isotopic spin of the nucleus T_ζ, defined as the sum of the projections of the isotopic spins of the neutrons and protons

$$T_\zeta = \sum_i \tau_{i\zeta} = \frac{1}{2}(Z - N),$$

is equal to half the neutron excess of the nucleus with sign reversed.

It is obvious that the projection T_ζ can be associated with any isotopic spin vector greater than $|T_\zeta|$; that is, a nucleus with an isotopic spin projection T_ζ can be characterized by states with isotopic spins which satisfy the inequality

$$T \geq |T_\zeta|.$$

Data on light nuclei (in which the isotopic spin is a good quantum number because of the small electric interaction) lead to the conclusion that the ground state of the nucleus is characterized by the smallest isotopic spin compatible with this inequality, that is to say, the isotopic spin of light nuclei in the ground state is given by

$$T_{\text{ground}} = |T_\zeta| = \frac{1}{2}(N - Z).$$

This rule is clearly connected with the nature of the interaction between the neutron and proton. We have seen that in an n-p system, of the two possible values of the isotopic spin, $T = 1$ (the state characteristic of two protons) and $T = 0$, the $T = 0$ state (deuteron) has a larger binding energy. (In the latter state the attraction forces do not bring about a real level.) This rule also applies to complex nuclei; for example, the Be^{10} nucleus and the C^{10} nucleus have

---

* The principal quantum number in the hydrogen atom is related to the energy of the level by $E_n = -(R_y/2n^2)$.

ground states with isotopic spin $T = 1$. The B^{10} nucleus has the same number of neutrons and protons and its isotopic spin in the ground state turns out to be $T = 0$. However, the level $T = 1$ in this nucleus, which is analogous to the level in Be^{10} and C^{10}, lies 1.74 Mev above the ground state.

It is possible to formulate a number of rules pertaining to ground state spins of nuclei. These rules determine the manner in which the angular momenta of the individual nucleons **j** are added to form the total moment of the nucleus.*

If the nucleus has an even number of protons and neutrons, that is, if the mass number A and the atomic number Z are both even, the spins add in such a way as to make the total momentum of the system zero

$$\sum_i \mathbf{j}_i = 0 \text{ (for nuclei with even } N \text{ and } Z).$$

Furthermore, if all the nucleons (beyond the closed shells) are in the same state, in nuclei in which A is odd, that is, in which either the number of protons or the number of neutrons is odd, the spins add in such a way as to make the total momentum of the system equal to the momentum of one particle.

For a long time this feature of nuclei has been used as a basis for the incorrect model of odd nuclei in which one (odd) particle moves in the field of all the other particles, which form a closed core.

Finally, in nuclei in which the number of protons and the number of neutrons are both odd, that is, in which A is even while Z is odd, and in which the neutrons and protons are in the same states (the same j and parity) the total momentum is equal to twice the momentum of one nucleon. As is well known there are only four such stable nuclei. These are H^2, Li^6, Be^{10} and N^{14}. Other nuclei of this type are radioactive.

The spin is connected with the magnetic moment of the nucleus. From analyses of experiments on atomic spectra it is well known that atomic magnetic moments (for example, those of the rare earths) are fairly well explained by the Russell-Saunders vector model. It is of interest to determine the degree to which the independent-particle vector model is able to explain the magnetic moments of light nuclei.

In determining the magnetic moment we may recall that the magnetic moment associated with a particle in the nucleus is made up of two parts: the inherent moment of the nucleon, parallel or anti-parallel to the spin of the nucleon, and the orbital moment, produced by the motion of the charged particle in its orbit— obviously, this part of the magnetic moment can arise only from protons.

In accordance with well-known formulas of electrodynamics the orbital moment of the proton is:

$$\boldsymbol{\mu} = \frac{e}{2c}\mathbf{r} \times \mathbf{v} = \frac{e}{2m_p c}\, \mathbf{r} \times \mathbf{p},$$

---

* At all times we shall be considering nucleons outside of closed shells, because the nucleons which form these shells make no contribution to the spin and magnetic moment of the nucleus.

where **v** is the velocity of the proton, m_p is its mass and **p** its momentum. Since **r** $\times$ **p** = 1 is the angular momentum of the proton,

$$\boldsymbol{\mu} = \frac{e\hbar}{2m_p c}\, \mathbf{l}.$$

The factor

$$\mu_N = \frac{e\hbar}{2m_p c}$$

is called the *nuclear magneton;* it is smaller than the *Bohr magneton*

$$\mu_0 = \frac{e\hbar}{2mc}$$

(m is the mass of the electron) by a factor m_p/m and is equal to $5.05 \cdot 10^{-24}$ erg·gauss^{-1}.

Nuclear magnetic moments are usually measured in nuclear magnetons so that in the following we will drop the factor μ_N from the formulas.

The spin magnetic moments of the proton and neutron have been measured experimentally. The magnetic moment of the free neutron is -1.91, while the magnetic moment of the free proton is 2.79.

In speaking of measurements of the magnetic moment of a nucleus one always understands the value of the magnetic moment averaged over the motion of particles in the nucleus. This mean magnetic moment of a nucleus is directed along the spin direction since this is the only preferred direction in the nucleus. Thus, we may write

$$\langle \boldsymbol{\mu} \rangle = g_j \mathbf{j};$$

the factor g_j is called the *gyromagnetic ratio.* We may note that the magnetic moment usually given in tables is the maximum value of the projection of $\langle \mu \rangle$ in the direction of the magnetic field, i. e., the quantity

$$\mu = g_j j.$$

The angular momentum of a nucleon **j** is made up of two parts

$$\mathbf{j} = \mathbf{l} + \mathbf{s},$$

and the magnetic moment of a nucleon can also be written as the sum

$$\boldsymbol{\mu} = g_l \mathbf{l} + g_s \mathbf{s}.$$

In this expression g_l and g_s are called, respectively, the orbital gyromagnetic ratio and the spin gyromagnetic ratio. It is apparent that these factors are equal for the proton and neutron. Specifically, assuming that the inherent value of s is $\pm^1/_2$, we have

for the proton $g_l = 1$ and
$$g_s = 5.58,$$

for the neutron $g_l = 0$ and
$$g_s = -3.82.$$

Our problem is to determine g_j for the proton and neutron, expressing this quantity in terms of g_l, g_s and j (or l).

We rewrite the expression for $\mathbf{\mu}$ in the form

$$\mathbf{\mu} = \tfrac{1}{2}(g_l + g_s)(\mathbf{l} + \mathbf{s}) + \tfrac{1}{2}(g_l - g_s)(\mathbf{l} - \mathbf{s}).$$

The average value of $\mathbf{\mu}$ must be along $\mathbf{j}$. In order to determine the gyromagnetic ratio we multiply this expression by $\mathbf{j} = \mathbf{l} + \mathbf{s}$. To compute the scalar product $\mathbf{\mu} \cdot \mathbf{j}$ we take the dot-product of the vector $\mathbf{\mu}$ and the vector $\mathbf{j}$. Thus

$$\mathbf{\mu} \cdot \mathbf{j} = g_j \mathbf{j}^2 = g_j j(j + 1) = \mu(j + 1);$$

here we have used the definition $\mu = g_j j$ and the identity $\mathbf{j}^2 = j(j + 1)$.

Using these expressions we find

$$g_j = \frac{1}{2}(g_l + g_s) + \frac{1}{2}(g_l - g_s)\frac{(l - s)(l + s + 1)}{j(j + 1)}.$$

Substituting $s = \tfrac{1}{2}$ and $j = l \pm \tfrac{1}{2}$, we have

$$g_j = g_l \pm \frac{g_s - g_l}{2l + 1}.$$

For the proton this expression yields

$$\mu = \left(1 - \frac{2.29}{j + 1}\right)j \quad \left(j = l - \frac{1}{2}\right),$$

$$\mu = \left(1 + \frac{2.29}{j}\right)j = j + 2.29 \quad \left(j = l + \frac{1}{2}\right)$$

while for the neutron we have

$$\mu = \frac{1.91}{j + 1}j \quad \left(j = l - \frac{1}{2}\right),$$

$$\mu = -\frac{1.91}{j}j = -1.91 \quad \left(j = l + \frac{1}{2}\right).$$

In Table 3 we present the values of g_j and $\mu = g_j j$ for the proton and neutron in various states.

TABLE 3

		$p_{1/2}$	$d_{3/2}$	$f_{5/2}$	$g_{7/2}$	$s_{1/2}$	$p_{3/2}$	$d_{5/2}$	$f_{7/2}$	$g_{9/2}$
Proton	g_j	-0.53	0.06	0.35	0.49	5.58	2.53	1.92	1.65	1.51
	μ	-0.26	0.13	0.86	1.72	2.79	3.79	4.79	5.79	6.79
Neutron	g_j	1.28	0.77	0.54	0.43	-3.82	-1.27	-0.76	-0.54	-0.42
	μ	0.64	1.15	1.36	1.49	-1.91	-1.91	-1.91	-1.91	-1.91

In computing the magnetic moment of a nucleus we must take the vector sum of the magnetic moments of all the nucleons and project it in the direction of the nuclear spin because, as in the case of a single nucleon, experimentally we can determine only the mean value of the nuclear magnetic moment. In this case nucleons in closed shells contribute nothing to the total magnetic moment and in the calculation we take account only of nucleons in unfilled levels of the nucleus.

If there is only an odd number of neutrons with the same j beyond the closed shells,

$$\langle \mathbf{\mu} \rangle_{\text{nucl.}} = \sum g_i \mathbf{j} = g_i \sum \mathbf{j} = g_i \mathbf{j}$$

by virtue of the addition rule for nucleon spins given above. The same relation applies for nuclei in which there are only protons beyond closed shells.

In these cases, both the spin and the magnetic moment can be described in terms of a model consisting of a single particle which moves in the field of the remaining nucleons. However, such a model is no longer valid in nuclei in which there are both neutrons and protons beyond the closed shells. In this case it is impossible to express the magnetic moment of a nucleus in terms of the spin since the gyromagnetic ratios of the neutron and proton are not the same. We shall carry out a calculation of the moments of a number of light nuclei in a later lecture.

Another characteristic electric property of the nucleus is the quadrupole moment. The quadrupole moment of a system of charges is given by the tensor

$$Q_{ik} = \sum_{\alpha} e_{\alpha} (3x_{i\alpha} x_{k\alpha} - \delta_{ik} r_{\alpha}^2),$$

where the summation extends over all particles (denoted by the index α). If the charged particles have the same charge e, this quantity can be taken out from the summation sign.

Usually the quadrupole moment is measured in barns (10^{-24} cm^2). In defining the quantity in this way we neglect the factor e and measure distance in units of 10^{-12} cm.

From the definition of the quadrupole moment it is apparent that the sign of the component Q_{zz} for a nucleus

$$Q_{zz} = \sum_{\alpha} (3z_{\alpha}^2 - r_{\alpha}^2)$$

(summed over all protons) is determined by the shape of the charged body. If the body is elongated along the z-axis, $Q_{zz} > 0$ (since the mean value of the z-coordinate is larger than the mean value of the square of the radius); if the body is compressed along this axis, $Q_{zz} < 0$.

In a quantized system, the quadrupole moment, like the magnetic moment, is related to the spin vector. To find the relation we make use of the following properties of the tensor Q_{ik}: a) it is symmetrical $Q_{ik} = Q_{ki}$, b) the sum of the diagonal elements is zero $Q_{xx} + Q_{yy} + Q_{zz} = 0$. These properties allow us to express Q_{ik} uniquely in terms of the spin vector (quantized) of the nucleus, **I**. Specifically,

$$Q_{ik} = A(I_i I_k + I_k I_i - \text{}^2\!/_3 \mathbf{I}^2 \delta_{ik}),$$

where A is a constant. (We may recall that the components of the spin vector do not commute.)

The quadrupole moment Q given in tables refers to the component Q_{zz} of this tensor in the state in which the projection of the spin on the z-axis is equal to the spin itself. Setting $I_z = I$ and $\mathbf{I}^2 = I(I + 1)$, we have

$$Q = \text{}^2\!/_3 A I(2I - 1).$$

It is apparent from this formula that when $I = 0$ or when $I = \text{}^1\!/_2$ the nucleus cannot have a quadrupole moment, since $Q = 0$ identically.

Using this same formula we can express the constant A in terms of Q and, making use of the value which is obtained, write the expression for the quadrupole moment tensor in the form

$$Q_{ik} = \frac{3Q}{2I(2I - 1)} \left[I_i I_k + I_k I_i - \frac{2}{3} I(I + 1)\delta_{ik} \right].$$

Structure of the Nucleus

(LIGHT NUCLEI)

We turn now to an analysis of the magnetic moments of nuclei, starting with light nuclei. We shall see that the measured values of the magnetic moments are in agreement with shell theory up to nuclei consisting of more than 20 particles.

n, p. We start with individual nucleons. The magnetic moments, measured in nuclear magnetons, are 2.79 for the proton and −1.91 for the neutron. It is obvious that speaking of states in one nucleon is meaningless; however, for consistency of notation we may speak of a single particle as being in a 1s state.

d. The deuteron is primarily in a state with moment $L = 0$. We will speak of the two particles as being in the $1s^2$ state, using the $1s^2$ designation for this state by analogy with atomic nomenclature for particles in a spherically symmetric field.

The magnetic moment for the $1s^2$ configuration should be equal to the sum of the moments of the two nucleons, i. e., 2.79 − 1.91 = 0.88. Actually the magnetic moment of the deuteron is 0.86. This may be taken as good agreement.

H³. The three nucleons which make up tritium can all be in 1s levels without contradicting the Pauli principle, since only two of them are identical. This structure results in a spin of $1/2$ for tritium. Thus, the H³ configuration may be designated by $1s^3$. The magnetic moment of this configuration is easily determined if one notes that there is one unfilled proton 1s state. It is clear that a system consisting of tritium plus a proton, in which this state is filled, will not have a magnetic moment. Hence, a system in a $1s^3$ configuration (one "hole" in the 1s shell) should have the same moment as the proton.* Again referring to this property the "hole" state is also sometimes designated by $1s^{-1}$. Thus, the theoretical value of the tritium magnetic moment is 2.79. The measured magnetic moment is 2.98. The discrepancy between theory and experiment is 0.2 magneton. Below we shall see that the discrepancy of 0.2 magneton is characteristic of the accuracy of the shell model with jj-coupling in magnetic moment problems.

He³. The next nucleus is the mirror image of tritium. Its spin is also $1/2$. Here we are dealing with the $1s^3$ configuration with the neutron site unfilled.

---

* The sign is the same since the sign of the magnetic moment is defined with respect to the spin direction.

Hence, the magnetic moment should be equal to the magnetic moment of the neutron, -1.91. Experimentally the value is found to be -2.13. As in tritium, the discrepancy is approximately 0.2 magneton, except that in this case the theoretical value is larger than the experimental value.

He⁴. The spin and magnetic moment are both zero. All four $1s$ states are filled; thus we arrive at the first closed shell ($1s^4$).

He⁵ and Li⁵. These nuclei do not exist in nature. They are unstable against separation into He⁴ and a nucleon. This is evidence of the fact that the $1s^4$ shell, which becomes filled at He⁴, is a closed system and that an additional nucleon cannot be added.

Data for proton and neutron scattering on He⁴ reveal a resonance associated with the intermediate nucleus He⁵ (or Li⁵), and indicate that the nucleon which is not part of the He⁴ occupies a $1p_{3/2}$ state.

Li⁶. The spin of this nucleus is 1. If, as might be expected from the simple scheme for filling shells, Li⁶ had the configuration He⁴$1p_{3/2}^2$, according to the empirical rule for forming moments in odd-odd nuclei its spin should be 3. This spin is actually observed in B¹⁰, which, as we shall see later, has the configuration $1p_{3/2}^{-2}$. It may be assumed that this empirical rule is not obeyed; in the Li⁶ nucleus the spins of the two nucleons add up in such a way as to form a total spin of 1.

This system could have a magnetic moment of 0.63, whereas the experimental value is $\mu(\text{Li}^6) = 0.82$. This discrepancy is not very great; it is 0.2 magneton, the discrepancy already noted for H³ and He³. However, a marked discrepancy arises when the quadrupole moment of the system is calculated. It is well known that the quadrupole moment of Li⁶ is no larger than 2% of the quadrupole moment of Li⁷ ($-4.6 \cdot 10^{-4}$ barns). This means that the quadrupole moment of Li⁶ is essentially zero. On the other hand, two nucleons in a $1p_{3/2}^2$ state should have a large quadrupole moment. It is apparent that in the Li⁶ nucleus we are dealing with an irregularity in the system for filling levels. This effect, which is frequently considered anomalous, is actually not strange. The same effect is encountered in electron shells in atoms. Here, the shells also filled in an irregular manner. It is well known, for example, that in the rare-earth region there is "competition" between the $4f$, $6s$ and $5d$ shells, and that in the uranium region the $5f$, $7s$ and $6d$ shells are filled together. Hence, one should not expect absolute regularity in the order of filling nuclear levels. The simplest hypothesis as to the structure of the Li⁶ nucleus is based on the assumption that both nucleons (for brevity we will frequently not bother to write the nucleons in the closed shells) occupy the new s-state, $2s_{1/2}$. The configuration $2s_{1/2}^2$ for Li⁶ is in agreement with the empirical rule of spin 1. The quadrupole moment of this state is obviously zero (spherical symmetry of the s-state!). The magnetic moment, as in the deuteron, equals the sum of the magnetic moments of the proton and neutron (the value is 0.88), and is also in good agreement with the experimental value.

If both nucleons were in the $1p_{1/2}$ state rather than $2s_{1/2}$, this would also give no quadrupole moment (the nucleon moment is $1/2$). This Li⁶ structure, however, would disturb the spin-orbit level scheme, and gives an incorrect value for the magnetic moment, namely 0.40.

Li⁷. The spin of this nucleus is $^3/_2$. The magnetic moment is 3.26. The nucleus consists of 3 nucleons—1 proton and 2 neutrons (beyond He⁴). These three particles should be in a $1p_{3/2}$ configuration, as is indicated by the spin value. We cannot use the simple considerations which have been used up to this point to compute the magnetic moment of this system. The wave function of a system of 3 particles is not unique even when the spin of the system is assigned. To determine the state uniquely it is necessary to determine the isotopic spin of the system. A general empirical rule indicates that the lowest state has the minimum isotopic spin. Hence, we assign to the ground state of Li⁷ an isotopic spin of $^1/_2$. The magnetic moment computed for this state ($1p_{3/2}^3$, $T = ^1/_2$, $I = ^3/_2$) is found to be 3.07, which is 0.2 smaller than the experimental value. These results seem to verify the validity of the above scheme.

Thus, in the Li⁷ nucleus we start filling the $p_{3/2}$ shell; this shell continues to be filled in subsequent nuclei.

Be⁹. The spin of this nucleus is $^3/_2$ and we shall assume that all 5 particles are in $1p_{3/2}$ states and that the spins add up in accordance with the rule which states that the total spin is equal to the moment of one particle. This configuration can also be written as a configuration in which three particles (2 protons and 1 neutron) are lacking for filling the C¹² shell. Thus, the Be⁹ configuration can be designated either by He⁴$1p_{3/2}^5$ or by C¹²$1p_{3/2}^{-3}$. The magnetic moment of this system also depends on its isotopic spin. Assuming that the isotopic spin of Be⁹ is $^1/_2$, we compute the magnetic moment. It turns out to be -1.14. The experimental value is -1.18. The agreement is rather good.

The Be⁹ nucleus is interesting in another respect. Using this nucleus, we can show the shortcomings of another model which is frequently discussed in the literature—namely, the α-particle model. From the point of view of the α-particle model the Be⁹ nucleus consists of 2 α-particles and 1 neutron. The magnetic moment due to 1 neutron in a state with $j = ^3/_2$ is either 1.15 (if this is the $d_{3/2}$ state) or -1.91 (if this is a $p_{3/2}$ state). Both of these values are in disagreement with experiment.

B¹⁰. The spin of B¹⁰ is 3; this value is in agreement with the rules and corresponds to 2 nucleons of different types in the $1p_{3/2}$ state. In the present case we are discussing unfilled configurations, that is, C¹²$1p_{3/2}^{-2}$. The calculated moment of this state is 1.88 and the experimental value is 1.80.

B¹¹. The spin of B¹¹ is $^3/_2$. We would expect a C¹²$1p_{3/2}^{-1}$ configuration for this nucleus and a corresponding magnetic moment, equal to the magnetic moment of the proton in the $p_{3/2}$ state. According to Table 3 of the previous lecture, this value is 3.79 and is in sharp disagreement with the experimental value 2.69. Thus, we again encounter an irregularity in filling the shells. It is reasonable to assume that the nucleons can avail themselves of the 2s state. Agreement with experiment can be reached only by assuming that 2 nucleons go into the new $2s_{1/2}$ state while the remaining 5 remain in the $1p_{3/2}$ state. Unfortunately, we cannot compare theory and experiment since the level is not assigned uniquely. The ambiguity is due to the fact that, although the total isotopic spin of the system is given, it is impossible to ascertain the way in which it is distributed between the two groups of particles. We can say only that the proposed scheme does not violate the magnitude of the quadrupole moment of this nucleus. Thus the following configuration can be assigned for B^{11}:

$$\mathrm{He}^4 1p_{1/2}^{5}\, 2s_{1/2}^{2}.$$

C^{12}. This nucleus has neither spin nor moment, and consists of two closed shells, $1s_{1/2}^{4}$ and $1p_{1/2}^{8}$. The filling of the $1p_{1/2}$ shell is not characterized by the same strong reduction in the binding energy as is the case for the $1s_{1/2}$ shell in He^4.

An idea of the degree to which a shell is closed can be obtained from the nucleon affinity in a nucleus, i. e., the energy evolved in adding a proton or neutron to this nucleus. It is found that the nucleon affinity in a nucleus with a closed shell is considerably smaller than in neighboring nuclei. With an increase in atomic weight, however, this effect is reduced because of irregularities in filling of the shells and the fact that all shell schemes become less reliable as the number of nucleons in the nucleus is increased. In He^4 the affinity is negative, since the nuclei He^5 and Li^5 do not exist. In C^{12} the neutron binding energy is 4.9 Mev, while that of the proton is 1.9 Mev. In the B^{10} nucleus the neutron and proton binding energies are, respectively, 11 and 9 Mev.

C^{13}. The spin of the C^{13} nucleus is $1/2$. This spin can be obtained if we add a nucleon in the $1p_{1/2}$ state or in the $2s_{1/2}$ state. In the first case the magnetic moment is 0.64 (cf. Table 3); in the second case it is -1.91. The experimental value is 0.70. The result determines the state uniquely as $C^{12}1p_{1/2}$.

N^{14}. The N^{14} nucleus corroborates the finding that the $1p_{1/2}$ state begins to be filled after C^{12}. The spin of N^{14} is 1, corresponding to a neutron and proton in the state $j = 1/2$. The magnetic moment of this state is 0.40, a value which is in agreement with the experimental value. Thus, there is no doubt that the configuration for this nucleus is $C^{12}1p_{1/2}^{2}$.

At this point we should note the extremely surprising long lifetime for the β-active nucleus C^{14}. Reasoning by analogy with N^{14}, this nucleus should have spin 0, corresponding to a configuration $C^{12}1p_{1/2}^{2}$ and the decay reaction $C^{14} \rightarrow N^{14}$ should be an allowed one. Actually, this reaction is strongly forbidden (the half-life is 5600 years, instead of several hours as is typical of an allowed decay). This anomaly indicates that there is a difference in the shell structure in C^{14} and N^{14}.

N^{15}. The N^{15} nucleus furnishes new evidence that the $1p_{1/2}$ level becomes filled. The spin is $1/2$ and the magnetic moment is -0.28. If this configuration is interpreted as one "hole" in a $1p_{1/2}$ configuration (lacking one proton for O^{16}): $O^{16}1p_{1/2}^{-1}$, the moment should be -0.24.

O^{16}. In this nucleus the $1p_{1/2}$ shell is filled. The structure is

$$1s_{1/2}^{4}1p_{1/2}^{8}1p_{1/2}^{4}.$$

Like the C^{12}, it is closed. The binding energy for a neutron in O^{16} is 4.1 Mev; the binding energy for a proton is 0.6 Mev. This quantity is considerably smaller than, say, in the O^{17} nucleus, in which the binding energy of a neutron is 7.9 Mev and that of a proton, 5.7 Mev.

O^{17}. The spin of the O^{17} nucleus is $5/2$. This means that a shell with $l = 2$ starts to be filled and we are encountering the $1d_{5/2}$ state for the first time. The neutron in the $d_{5/2}$ state has a magnetic moment of -1.91, which is in excellent agreement with the value -1.89 measured experimentally.

F^{19}. Contrary to our simple assumption, the spin of F^{19} is not $5/2$, as would be the case if the $1d_{5/2}$ state were filled; rather, it is $1/2$. This value apparently means that $2s_{1/2}$ states enter, displacing the $1d_{5/2}$ state.

The magnetic moment of the system $Ne^{20}(2s_{1/2})^{-1}$ should be equal to the moment of the proton, 2.79. Experimentally the value is found to be 2.63.

Ne^{20}. In accordance with what has been said above this nucleus consists of four closed configurations:

$$1s_{1/2}^4\ 1p_{3/2}^8\ 1p_{1/2}^4\ 2s_{1/2}^4.$$

Ne^{21}. The spin of this nucleus is $3/2$, indicating a $2p_{3/2}$ configuration. In this case (Ne^{20} plus a $p_{3/2}$ neutron) the magnetic moment should be -1.91. There are indications that the magnetic moment of Ne^{21} is negative. Further study of this nucleus will be of great interest.

Na^{22}. This is a radioactive nucleus in which both the spin (3) and the magnetic moment (1.75) are known. It is most reasonable to assume that the two nucleons (beyond Ne^{20}) are in $2p_{3/2}$ states. Thus, the Na^{22} configuration is very similar to a B^{10} configuration, in which there are two "holes" in the $1p_{3/2}$ level. The spin of this system should be 3, and the magnetic moment, as we have seen in the B^{10} case, 1.88. We may recall that the magnetic moment of Na^{22} (experimental) is 1.79, that is, almost equal to the B^{10} moment; these results tend to support our picture of the structure of these two nuclei.

Mg^{25}. This is the last nucleus whose structure and configuration can be analyzed without introducing a large number of assumptions. The spin is $5/2$. The magnetic moment is -0.85. These quantities are reasonably well explained if we assume a configuration $O^{16}d_{5/2}^9 \equiv Si^{28}d_{5/2}^{-3}$ in Mg^{25}. With $T = 1/2$ this configuration has a magnetic moment which agrees with experiment. It is curious that in this case even those $2s_{1/2}^4$ levels which are being filled seem to vanish.

Further comparison of the theoretical values of the magnetic moments with the experimental values is not fruitful because the large numbers of particles in the shells mean that the computations must be ambiguous. However, the success of the independent particle model in describing light nuclei is apparent even from the cases which have been considered. To make these considerations clearer we have collected all the data in Table 4, in which the values of the spins and magnetic moments of the light nuclei are tabulated.

The independent particle scheme makes it possible to understand the magnetic moment and spins of the light nuclei and to develop a number of characteristics of those nuclear reactions in which excited states and ground states of the nuclei are both of importance. A direct measurement of the spin and magnetic moment of excited states of nuclei is beyond the capabilities of present-day

TABLE 4

Nucleus	Spin	Magnetic moment	Configuration	Magnetic moment (theoretical)
		Measured values		
n	$1/2$	-1.91	$1s_{1/2}$	
H^1	$1/2$	2.79	$1s_{1/2}$	
H^2	1	0.86	$1s_{1/2}^2$	0.88
H^3	$1/2$	2.98	$1s_{1/2}^3$	2.79
H^3	$1/2$	-2.13	$1s_{1/2}^3$	-1.91
He^4	Closed shell		$1s_{1/2}^4$	
Li^6	1	0.82	$He^4\, 2s_{1/2}^2$	0.88
Li^7	$3/2$	3.26	$He^4\, 1p_{3/2}^3$	3.07
Be^9	$3/2$	-1.17	$He^4\, 1p_{3/2}^5$	-1.14
B^{10}	3	1.80	$He^4\, 1p_{3/2}^6$	1.88
B^{11}	$3/2$	2.69	$He^4\, 1p_{3/2}^5\, 2s_{1/2}^2$	Not unique
C^{12}	Closed shell		$1s_{1/2}^4\, 1p_{3/2}^8$	
C^{13}	$1/2$	0.70	$C^{12}\, 1p_{1/2}$	0.64
N^{14}	1	0.40	$C^{12}p_{1/2}^2$	0.40
N^{15}	$1/2$	-0.28	$C^{12}p_{1/2}^3$	-0.24
O^{16}	Closed shell		$1s_{1/2}^4\, 1p_{3/2}^8\, 1p_{1/2}^4$	
O^{17}	$5/2$	-1.89	$O^{16}d_{5/2}$	-1.91
F^{19}	$1/2$	2.63	$O^{16}2s_{1/2}^3$	2.79
Ne^{20}	Closed shell		$1s_{1/2}^4\, 1p_{3/2}^8\, 1p_{1/2}^4\, 2s_{1/2}^4$	
Na^{21}	$3/2$	<0	$Ne^{20}\, 2p_{3/2}$	-1.91
Na^{22}	3	1.75	$Ne^{20}\, 2p_{3/2}^2$	1.88
Mg^{25}	$5/2$	-0.85	$O^{16}\, 1d_{5/2}^9$	-1.06
Si^{28}	Closed shell		$1s_{1/2}^4\, 1p_{3/2}^8\, 1p_{1/2}^4\, 1d_{5/2}^{12}$	

experimental techniques; hence, the identification of these states must be carried out from studies of the selection rules and the angular distributions of the various particles. Analyses of this type have already yielded a large amount of information as to the levels in light nuclei, and this information is in agreement with the configuration scheme considered above. We present several examples below.

The light nuclei H^2, H^3 and He^3 have no low-lying excited states and the particle interaction in these nuclei is so small that the "potential wells" have only one level, corresponding to the ground state.

Similarly, there are no levels in α-particles. However, there is evidence that an unstable level does exist in α-particles; this level can decay into tritium plus a neutron. It is believed that this level appears in the reaction

$$H^3 + p \rightarrow He^3 + n.$$

A very wide resonance is observed in this reaction at an energy of 2.6 Mev (in the c.m. system). This resonance can be interpreted as the appearance of a level with an energy of approximately 22 Mev and a width of approximately 1 Mev in the α-particle. Such a level arises if one of the nucleons in the α-

particle makes a transition from the $1s_{1/2}$ state to the $1p_{3/2}$ state. In this case the total spin of the system is determined by the vector sum $1/2 + 3/2$, and is equal to either 2 or 1 (states which are obviously odd). It is possible that an experiment can reveal the two states 2 and 1 separated by approximately 1 Mev. We may note that the absence of a stable level in the α-particle makes it possible to draw certain conclusions as to the stability of H^4. One proton and three neutrons cannot all be in the $1s_{1/2}$ state because of the Pauli principle. Hence, if H^4 exists, one of the neutrons must be in the $1p_{3/2}$ state, and the H^4 nucleus must be similar to an excited state of He^4. However, there are no excited states in He^4 which are stable against emission of a nucleon; consequently, there are none in the H^4 system. It may also be noted that there is no guarantee that the excited state of He^4 has an analog in the H^4 system. The H^4 system has an isotopic spin of at least 1 (projection $T_\zeta = 1$). The excited state of He^4 can also have an isotopic spin of zero.

An interesting effect is observed in the He^5 system. It has already been noted that in scattering of neutrons by nuclei there is a resonance corresponding to the formation of He^5 in the intermediate state. This level is unstable against decay to $He^4 + p$ (it has an energy of 1 Mev), and corresponds to a $1p_{3/2}$ neutron configuration.

The same pattern is observed in the scattering of protons on helium, which leads to the production of the Li^5 nucleus in the intermediate state. In this case, because of Coulomb forces, the level lies somewhat higher than in He^5 (1.8 Mev).

There is one other level in He^5 which is of interest. It is well known that the reaction

$$H^3 + H^2 \rightarrow He^4 + n$$

has a very sharp resonance at 15 kev (in the c.m. system). The value of this maximum corresponds to the maximum possible cross section in a system of spin $3/2$. The surprising thing about this reaction is that the resonance is very narrow (width of approximately 40 kev) in spite of the fact that the neutrons in this reaction have high energy. The narrowness of this resonance indicates a comparatively high stability for the nucleus at this high level (16.6 Mev above the ground state of He^5), which is strange. Actually, it is easy to explain this reaction if it is assumed that two particles in the He^5 system are excited to the $1p_{3/2}$ level. Since the excitation of He^4 requires an energy of approximately 20 Mev, the appearance of a level at this height is completely reasonable. On the other hand, the decay of this excited He^5 nucleus into an α-particle and neutron should be characterized by transitions of two particles, one neutron—the emitted free neutron (continuous spectrum)—and another nucleon, to the $1s_{1/2}$ ground state. It is reasonable that a transition of this type will have a considerably smaller probability than the single-particle transitions which are encountered in the majority of reactions. This situation explains the relative stability of this state.

A similar effect is observed in the Li^5 system which is produced in the reaction

$$He^3 + He^2 \rightarrow He^4 + H^1.$$

Here the resonance lies at a somewhat higher energy (260 kev) and, correspondingly, has a smaller magnitude. [As is well known, the maximum cross section is proportional to $\lambda^2 \sim 1/E$.]

In interpreting nuclear reactions one must also take account of the selection rules. The best known example of this kind appears in the decay of the Be^8 nucleus into two α-particles. As is well known, the two α-particles can be only in symmetric states (as for any two identical particles with no spin). This requirement means that they can separate only in **s, d,** or **f** states (l even). Hence, the Be^8 nucleus can decay into two α-particles only from those states which have even spin and positive parity (isotopic spin zero). This selection rule is apparent in the $Li^7 + p$ reaction. This reaction can occur with the emission of two α-particles,

$$Li^7 + p \rightarrow 2\alpha,$$

or, by the above reasoning, with the emission of γ-quanta. In this case the reaction occurs in two stages:

$$Li^7 + p \rightarrow Be^8 + \gamma.$$
$$\ \llcorner\!\rightarrow 2\alpha.$$

An interesting feature of this reaction is the fact that the γ-ray yield curve has a maximum at a proton energy of 441 kev (which can be associated with the Be^8 level), whereas the α-particle yield curve has no maximum at this energy. This means that the Be^8 nucleus, in the state which is formed at this proton energy, cannot divide into two α-particles. This forbiddenness is explained by the fact that the Be^8 level in this reaction (excitation energy of 17.63 Mev) has spin 1 and is odd. This level cannot decay into two α-particles, and the excitation can be dissipated only by radiation.

The isotopic spin selection rule appears in an interesting way in reactions. If, for example, we consider the inelastic scattering of deuterons or α-particles on nuclei, in which the nucleus is left in an excited state, the isotopic spin is zero for both the deuteron and α-particle, and we can excite only those levels which have the same isotopic spin as the ground state. For example, in the elastic scattering of deuterons on B^{10} we excite levels in B^{10} with energies of 0.72, 2.15 and 3.58 Mev. In inelastic proton scattering, however, in addition to these levels we excite the level with an energy 1.74 Mev. Since the isotopic spin of B^{10} is zero, it follows unambiguously that the isotopic spin of the level at 1.74 Mev must be unity. This is the level which is similar to the ground state of B^{10} and C^{10}.

The simplest manifestation of isotopic invariance occurs in the mirror nuclei Be^7-Li^7, C^{13}-N^{13}, O^{17}-F^{17}, and so on. In these nuclei each level of one nucleus corresponds to a level of another and thus their spectra differ only by the Coulomb field (and mass differences due to the difference in the mass of the proton and neutron).

A comparison of isobaric nuclei is sometimes desirable for determining the isotopic spin of a level. For example, consider the nuclei N^{16} and O^{16}. The N^{16} nucleus has a projected isotopic spin $T_\zeta = -1$, while the O^{16} nucleus has a projection $T_\zeta = 0$. This is apparent from the fact that the neutron excess $N - Z$ in N^{16} is 2, while this same quantity is zero in O^{16}. Thus, the levels in N^{16} can have isotopic spin $T = 1, 2, \ldots$, while the O^{16} nucleus can have only levels with $T = 0$. All levels of the N^{16} nucleus should be encountered in the spectrum of the O^{16} nucleus, but not all the O^{16} levels are included in the N^{16} spectrum. Those levels with $T = 0$ will not appear in the N^{16} spectrum. The transition from a level in the O^{16} nucleus to a corresponding level in the N^{16} nucleus is accomplished by replacing one proton by a neutron. Hence, the N^{16} nucleus in this state will have a mass differing from the mass of the O^{16} nucleus in the corresponding state by the Coulomb energy of the missing proton minus the difference of the mass of the neutron and proton. In light nuclei this quantity is small.* It is equal to zero (more precisely, 15 kev) in the pair H^3, He^3 and increases with increasing nuclear charge. A comparison of the spectra of neighboring nuclei is also desirable for identification of the levels. It is possible, in particular, to state that O^{16} in the ground state has an isotopic spin of zero (O^{16} in the ground state has a mass which is smaller than the N^{16} mass).

This type of analysis can also be carried out for other isobars. Supplementing this information with other data on nuclear reactions we are now in a position to describe a large number of the lowest excited states of light nuclei.

---

* As is customary in nuclear physics, we speak of the mass of the neutral atom rather than the mass of the nuclei.

LECTURE SIX

Structure of the Nucleus

(HEAVY NUCLEI)

*I*n the light-nucleus region we have assumed that the neutron and proton are "isotopically" similar; in the heavy-nucleus region, in which an important role is played by Coulomb forces, this assumption is no longer valid. Whereas the properties of light nuclei depend on the total number of nucleons in a shell, the properties of heavy nuclei depend on the number of neutrons, N, and the number of protons, Z, individually.

The explicit dependence of nuclear properties on N and Z is manifest in the existence of so-called "magic numbers" and corresponding "magic nuclei."

Magic nuclei are nuclei in which either the number of protons or the number of neutrons corresponds to a filled shell. The properties of magic nuclei are rather different from those of other nuclei.

It is now generally believed that the numbers 50, 82, 126 and 152 are *magic numbers*. Other magic numbers which are encountered in the literature refer either to extremely light nuclei, in which these numbers have little significance, or to numbers whose existence is open to doubt.

The best example of a magic nucleus is the lead isotope of atomic weight 208. This nucleus is, so to speak, a doubly magic nucleus, since it contains a magic number of protons (82) and a magic number of neutrons (126). Pb^{208} exhibits properties which are characteristic of a closed system. Whereas the first excited levels in ordinary heavy nuclei are never higher than 200 kev, the excitation of Pb^{208} requires an energy of 2.6 Mev. Furthermore, the binding energy for an additional nucleon is anomalously small. Neutron capture is usually characterized by the excitation of the resulting nucleus to about 7 Mev; neutron capture in Pb^{208} leads to an excitation of only 4 Mev.

Probably the best way of determining the magic nuclei is an investigation of the affinity of nuclei for α-particles (i. e., the energy released in capture of an α-particle by a nucleus). The dependence of the affinity for α-particles on the number of nucleons in the nucleus is a more reliable characteristic than the affinity for neutrons or protons, which depends on the number of particles in an irregular way, changing discontinuously in going from even to odd elements. Studies of the systematics of α-active elements and reactions involving the emission of α-particles are extremely useful in this connection.

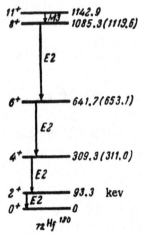

Fig. 6. Spectrum of the nucleus $_{72}\mathrm{Hf}^{180}$ with an even number of protons and neutrons (from A. Bohr, *Rotational States of Atomic Nuclei*, Copenhagen, 1955). The energies corresponding to the formula for rotational levels are given in parentheses. The level 11 + which appears at the highest energy is not rotational in character. The arrows indicate observed electromagnetic transitions (*E2* electric quadrupole, *M3* magnetic octupole). It is apparent that the quadrupole transitions occur between rotational levels (radiation of a charged drop). It is also clear that the last level is not a rotational level.

It is unfortunate that, having established the magic numbers, there is still very little we can say regarding the nucleon states. Many more experiments will be required before the pattern of states for heavy nuclei becomes in any way complete. The level scheme and the order of filling which frequently appear in the literature are based on a simplified model of the potential well and an idealized filling system which, as is apparent from the light-nucleus examples, does not correspond to reality.

Moreover, experiment indicates that in heavy nuclei the model of a particle moving in a central field requires essential modification because the original assumption as to spherical symmetry is no longer valid.

It is undoubtedly true that magic nuclei are spherical in shape. However, this shape is extremely unstable, especially for heavy nuclei; the Coulomb forces, which tend to elongate the nucleus, and the surface tension forces, which tend to reduce the surface in heavy nuclei, almost completely cancel each other. Hence, even under the effect of relatively weak forces the nucleus becomes a sphere.

The instability of heavy nuclei is very well illustrated by nuclear fission. With comparatively small excitation (approximately 6 Mev) the U^{236} nucleus, which is formed as a result of thermal neutron capture, becomes unstable and somewhat elongated; eventually this situation leads to its division into two fragments. In the nuclei U^{238}, Pu^{240}, etc., the instability is even larger—these nuclei fission spontaneously, being found in the ground state. Because of this instability many heavy nuclei in the ground state are elongated rather than

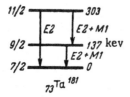

Fig. 7. Spectrum of the nucleus $_{73}Ta^{181}$ with an odd number of protons and an even number of neutrons (same reference as Fig. 6). The spin of the ground state corresponds to a nucleon spin projection $\Omega = {}^1/_2$. The first rotational level has an energy which almost corresponds to the same moment of inertia as the neighboring nucleus $_{72}Hf^{180}$. The second level is also described by the rotational level formula. The energy ratio $E_{11/_2}:E_{9/_2} = 2.21 \pm 0.02$ is in agreement with the theoretical value 2.22. The rotational nature of the levels is corroborated by the observed quadrupole transitions, which, however, compete here with the magnetic transitions—dipole transitions associated with a change of Ω.

spherical. The elongation of the nucleus is related to the interaction between the surface and the nucleons outside closed shells. Without an analysis of the experimental data it is impossible to say in which nuclei the elongation becomes sufficiently large to make the nucleus nonspherical. The light nuclei are undoubtedly spherical—this observation is based on the success of shell theory, which is based on states in a field of spherical symmetry. From what has been indicated above it is clear that a particularly marked nonsphericity is to be expected in nuclei at the end of the periodic table.

The basic ideas concerning nonspherical nuclei have been most completely described by A. Bohr.* We now examine the properties of nonspherical nuclei.

The most important distinction between nonspherical and spherical nuclei is that the former can have rotational levels. According to the basic ideas of quantum mechanics, it is meaningless to discuss rotation of a spherical nucleus. A spherically symmetric system, i. e., a system whose properties are independent of angle, cannot have a rotational energy spectrum. The concept of rotation in such a system is meaningless; similarly, it is meaningless to speak of rotation in a spherically symmetric nucleus. However, in an elongated nucleus the concept of rotation is meaningful. If, for simplicity, it is assumed that the nucleus is an elongated ellipsoid of rotation, the nucleus can rotate about an axis perpendicular to the axis of symmetry. Rotation about the symmetry axis of the nucleus, however, is meaningless for the reasons given above.

A nonspherical nucleus is characterized by the state of the "quiescent" nucleus and by the moment of inertia about the axis perpendicular to the symmetry axis of the nucleus.

* We refer the reader to the collection "Problems of Contemporary Physics" No. 9, Foreign Lit. Press, 1955, in which there are translations of the two basic papers on nonspherical nuclei. (Cf. also A. Bohr, *Rotational States of Atomic Nuclei*, Copenhagen, 1955.)

The concept of angular momentum for the individual nucleons loses significance in the field of a nonspherical nucleus because this quantity is a good quantum number only in a spherically symmetric field. The nucleon state in this field must be classified by analogy with a diatomic molecule.

In a field of axial symmetry the quantity which is conserved is not the angular momentum, but rather its projection on the symmetry axis. The projection of the angular momentum of the nucleon is made up of the projection of its orbital moment and the projection of its spin; it is not meaningful to speak of the total moment of a nucleon j in a field of axial symmetry.

We shall designate the sum of the projections of the nucleon angular momenta on the nuclear axis by the symbol Ω.

Let us consider a rotating nucleus. Rotation of a nucleus is characterized by the angular momentum $\mathbf{K}$ which, as we have seen, has no component along the symmetry axis of the nucleus. If we use the symbol $\mathbf{n}$ to denote a unit vector along the symmetry axis of the nucleus, the total spin of the nucleus can be written in the form:

$$\mathbf{I} = \Omega\mathbf{n} + \mathbf{K}.$$

The vector $\mathbf{K}$ is perpendicular to the symmetry axis, and thus

$$\mathbf{K}\cdot\mathbf{n} = 0.$$

From this it follows that $\mathbf{I}\cdot\mathbf{n} = \Omega$, i. e., the projection of the nuclear spin on the symmetry axis is always equal to Ω.

From the obvious consideration that the projection of a vector cannot exceed its length, it follows that for a given Ω the possible values of a nuclear spin are

$$I = \Omega, \Omega + 1, \Omega + 2, \ldots, \text{etc.}$$

Special attention is required for a nucleus in which $\Omega = 0$, that is, a nucleus with zero spin. As the theory indicates, such nuclei are similar to diatomic molecules with identical spinless nuclei; hence the angular momentum can only be an even number—$0, 2, 4, \ldots$.

A study of nuclear spectra with the purpose of identifying rotational levels leads to interesting results. The most striking rotational spectra appear in the heavy α-active nuclei. The rotational levels are best seen in even-even nuclei, in which the excitation of individual nucleons requires a comparatively large energy and the nucleon excitation spectrum is not superposed on the rotational spectrum. An example of such a spectrum is given in Fig. 6.

It is also possible to separate out rotational levels in odd nuclei, using the features described below. In Fig. 7 we show the rotational spectrum in the odd nucleus $_{73}\text{Ta}^{181}$.

It should be kept in mind that experimentally it is difficult to separate rotational levels with high moments. However, at the present time, rotational levels with moments of 8 have been observed.

The energy associated with the rotational levels of a nucleus is given by the well-known formula

$$E = \frac{M^2}{2J},$$

where M is the angular momentum and J is the moment of inertia of the nucleus. In a quantized system $M^2 = \hbar^2 I(I + 1)$, and the energy of a level with moment I is given by

$$E_I = \frac{\hbar^2}{2J} I(I + 1).$$

J, the moment of inertia of a nucleus, has little in common with the moment of inertia of a solid ellipsoid. In the case of spherical symmetry the latter is a finite quantity—the moment of inertia of a sphere. However, rotation of a spherical nucleus, as we have seen, is not physically meaningful, and the most important property of the moment of inertia J is that $J \to 0$ as the nuclear shape becomes spherical.

Thus, nuclear rotation cannot be described in terms of the rotation of a solid body. A theoretical analysis of a nonspherical nucleus indicates that rotation can be described approximately by equations which are similar to the hydrodynamic equations of the potential motion of an ideal liquid in a rotating asymmetrical shell. However, the suitability of this approximation is still not completely established.

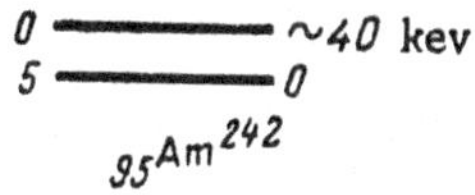

Fig. 8. Probable scheme for the first levels of $_{95}Am^{242}$ identified by β-decay of curium (from the data of S. A. Baranov and K. N. Shlyagin). The nucleus $_{95}Am^{242}$ has an odd number of protons as well as an odd number of neutrons. The spins of the levels arise as a result of addition and subtraction of the projections of the moments of both nucleons, each of which is $^5/_2$. It is apparent from the scheme that the addition of the projections is to be associated with the lower level.

The solution of this problem is characterized by precisely the same features as those indicated above. The usual conditions of hydrodynamics require that the normal component of the velocity of a liquid be zero at a boundary defined by a fixed wall. If the wall is not fixed, but moves with a given velocity, the boundary conditions require that the normal (to the wall) component of the velocity be the same as the normal component of the wall velocity.

We consider a liquid enclosed in a spherical container rotating about an axis through its center. It is apparent that in such motion the velocity of each point of the surface is tangential and that the normal component of the wall velocity is zero everywhere.

Only a liquid at rest can satisfy the hydrodynamic equations for potential motion in such a container, since $\mathbf{v} = 0$ satisfies both the equations and the boundary conditions. In a nonspherical container, however, the walls have a normal velocity component and the liquid is entrained by the walls; moreover, the kinetic energy for a given angular velocity of the shell becomes greater as the container becomes less spherical.

We shall use as a model a nucleus in the shape of an ellipsoid. The semi-axes of the ellipsoid are denoted by c (symmetry axis) and a. It can be shown that the kinetic energy associated with potential rotation of a liquid ellipsoid is

$$\frac{M}{10} \frac{(c^2 - a^2)^2}{c^2 + a^2} \omega_0^2,$$

where ω_0 is the angular velocity of rotation of the ellipsoid. Comparing this expression with the general expression for the kinetic energy of a rotating body,

$$\tfrac{1}{2} J\omega_0^2,$$

we have

$$J = \frac{M}{5} \frac{(c^2 - a^2)^2}{c^2 + a^2}.$$

We note that the moment of inertia of a solid ellipsoid is

$$J_0 = \frac{M}{5} (c^2 + a^2),$$

and that the ratio of these quantities is

$$\frac{J}{J_0} = \left(\frac{c^2 - a^2}{c^2 + a^2} \right)^2.$$

This ratio approaches zero as $(c/a) \to 1$ (sphere), and unity when $(c/a) \to \infty$.

The volume of the ellipsoid is

$$V = {}^4/_3 \pi a^2 c.$$

The mean radius of the nucleus, which is the quantity generally considered in nuclear physics, is obviously

$$R = (a^2 c)^{1/3}.$$

We now consider the rotational energy levels. In a nucleus without spin (even-even) the energy levels are given by the expressions:

$$E_I = \frac{I(I + 1)}{2J} \hbar^2; \quad E_0 = 0; \quad E_2 = \frac{3\hbar^2}{J}; \quad E_4 = \frac{10\hbar^2}{J};$$

$$E_6 = \frac{21\hbar^2}{J}, \text{ etc.}$$

We compare these expressions with the experimental energy levels for heavy nuclei. We must first verify the interval rule—the ratio of the energy differences of the levels—which should be

$$(E_2 - E_0):(E_4 - E_2):(E_6 - E_4):\ldots = 3:7:11:\ldots,$$

or, computing the energy from the ground state of the nucleus,

$$E_2:E_4:E_6:E_8:\ldots = 1:{}^{10}/_3:7:12:\ldots.$$

This is in excellent agreement with the levels of two groups of even-even nuclei—in the rare earth region and in the region of heavy α-active nuclei. It is extremely interesting that there are no rotational levels in Pb^{208}, which is a doubly-magic spherical nucleus; it is only as one goes away from lead toward the heavier nuclei that rotational levels arise and the ratio $E_4:E_2$ approaches the theoretical value 3.3. The distance to the first excited level is approximately 80 kev in the rare earth region and approximately 40 kev in α-active nuclei. It is thus possible to determine the moments of inertia. In the rare earths $J = 1.9 \cdot 10^{-47}\,\mathrm{g \cdot cm^2}$, while in the heavy α-active nuclei $J = 3.8 \cdot 10^{-47}\,\mathrm{g \cdot cm^2}$.

The interval rule for the levels is best illustrated by examples. One of these is the U^{234} spectrum, which is measured from the energy of the α-particles of Pu^{238} decay. Here there are three rotational levels with energies of 43, 143 and 297 kev with spins and parity, respectively, 2^+, 4^+ and 6^+. The ratio of the energy levels is $1:3.3:6.9$, in strict agreement with the theoretical values.

A second example—the spectrum of the rare earth element $_{72}Hf^{180}$ (Fig. 6)—appears in the decay of the isomeric state (5.5 hours). Here, four levels are apparent—2^+, 4^+, 6^+ and 8^+; these have energies of 93, 309, 642 and 1085 kev, which are in the ratio $1:3.3:6.9:11.7$. The spins and parities of the levels are such that all transitions between these levels can be identified as electric quadrupole transitions.

The expression for distances between levels is different in odd nuclei. Since the spin in these nuclei is not zero in the ground state, the levels are determined by the expression

$$E_I = \frac{\hbar^2}{2J} I(I + 1),$$

where $I = I_0, I_0 + 1, I_0 + 2$, and so on ($I_0$ is the spin of the ground state of the nucleus); an example of such a spectrum is given in Fig. 7.

It is necessary to have one other experimental quantity to determine the elongation of the ellipsoid, since the shape of the ellipsoid is specified by two quantities, a and c. This can be done by introducing other data on nonspherical nuclei. A comparison of this type is carried out below.

We now consider the individual nucleon states in a nonspherical nucleus.

First of all, a good quantum number for a nucleon in an axially symmetric field is λ, the projection of the nucleon orbital angular momentum on the ellipsoid axis. Since the spin of the nucleon can have a projection equal to $\pm 1/2$ along the nuclear axis, the total projection of the nucleon angular momentum ω can be either $\lambda + 1/2$ or $\lambda - 1/2$. We may note a single obvious exception. If $\lambda = 0$, ω can assume only one value: $\omega = 1/2$.

In the ellipsoidal field the nucleon levels are designated in spectroscopy by the value of the projection λ. Just as the values $l = 0, 1, 2, 3, \ldots$ are denoted by the symbols $s, p, d, \ldots$, the levels with $\lambda = 0, 1, 2, 3, \ldots$ are denoted by the symbols $\sigma, \pi, \Delta, \ldots$.

Inasmuch as a nucleon in a nonspherical nucleus does not have a definite angular momentum vector, we cannot use vector addition for the momenta. In a system of several nucleons we can only add the projections of the momenta on the nuclear axis. In view of this situation, the scheme for filling levels in a nonspherical nucleus proves to be completely different from that which applies for the central field in light nuclei.

The chief difference is in the following. Particle levels with momentum j in a central field are highly degenerate. Each value of j corresponds to $2(2j + 1)$ states with the same energy, differing either in charge or projection on an arbitrary axis. Thus, in a state with given j there can be $2j + 1$ neutrons and as many protons.

When a particle is in an axially symmetric field the degeneracy is considerably smaller. In a level characterized by a given projection we can have two nucleons with $\pm \omega$, respectively. In this case, because protons and neutrons occupy different levels in heavy nuclei, the charge degeneracy (isotopic spin) does not exist. Hence, each pair of identical nucleons forms a closed shell which has neither spin nor magnetic moment. On the other hand, in light nuclei, we need $2(2j + 1)$ nucleons to form a closed shell. Hence, to a considerable degree the properties of the heavy nuclei are determined by the properties of the last odd particle, a fact which leads to a somewhat paradoxical situation: heavy nuclei, with a large number of particles, can be described in terms of the single-particle model, whereas, as we have seen, this model is unsuitable for light nuclei.

If there are two different particles beyond the closed shells—one proton and one neutron—these nucleons occupy different levels, and the behavior of the nucleus becomes more complicated. If the interaction between the nucleons is small the energy of the level depends to some extent on the sign of the projection of the angular momentum of the nucleus. Thus, one would expect the appearance of two levels, lying close together, with spins equal, respectively, to the sum of the projections of the moments of the two nucleons and the difference. Such a spectrum has been observed, for example, in the nucleus Am^{242} * (Fig. 8); the lower level apparently has spin 5, while the spin of the upper level is zero. This picture corresponds to two nucleons, each with a spin projection of $^5/_2$. The lower level corresponds to the sum of the projections, while the upper level corresponds to the difference. The spacing between the levels is approximately 40 kev; this value gives an idea of the magnitude of the interaction between the nucleons.

We start our analysis of the properties of heavy nuclei by considering spin. Disregarding odd-odd nuclei for the moment, we may say that the spin of the nucleus is determined by the state of the odd nucleon. The nuclear spin coincides with ω, the projection of the total angular momentum of the nucleon on the nuclear axis. The projection of the orbital moment λ is either $\omega - ^1/_2$ or $\omega + ^1/_2$.

---

* Baranov and Shlyagin, Report to the Conference on the Peaceful Uses of Atomic Energy, Academy of Sciences USSR, July 1955.

Since each pair of identical nucleons comprises a closed system, in the region of heavy nuclei the spins of neighboring nuclei follow each other in an irregular way.

The magnetic moment of a heavy nucleus is made up of two parts. In addition to the magnetic moment due to the individual nucleons in the "quiescent" nucleus, there is an additional magnetic moment due to the rotation of the nucleus as a whole. The ratio of the magnetic moment to the mechanical moment due to the rotation of the nucleus as a whole is given by $Ze/2Amc$, where Ze is the charge and Am the mass of the nucleus. In nuclear magnetons the gyro-magnetic ratio is given by

$$g_k = \frac{Z}{A}.$$

Thus, the total moment $\boldsymbol{\mu}$ of the nucleus is the sum:

$$\boldsymbol{\mu} = \mu_0 \mathbf{n} + \frac{Z}{A}\mathbf{K},$$

where $\mathbf{n}$, as before, is a unit vector along the nuclear axis, $\mathbf{K}$ is the angular momentum of the nucleus and μ_0 is the projection of the magnetic moment of the odd nucleon on the symmetry axis.

We now determine the projection of the magnetic moment in the spin direction μ. As we know, it is this quantity which is called the *magnetic moment of the nucleus*. By definition:

$$\boldsymbol{\mu} = \mu \frac{\mathbf{I}}{I},$$

whence

$$\mu = \frac{\boldsymbol{\mu} \cdot \mathbf{I}}{I+1}.$$

To compute the scalar product $\boldsymbol{\mu} \cdot \mathbf{I}$, we multiply the expression for $\boldsymbol{\mu}$ by the spin of the nucleus $\mathbf{I}$:

$$\boldsymbol{\mu} \cdot \mathbf{I} = \mu_0 \mathbf{n} \cdot \mathbf{I} + \frac{Z}{A}\mathbf{K} \cdot \mathbf{I}.$$

The nuclear spin $\mathbf{I}$ consists of two terms: $\mathbf{I} = \Omega \mathbf{n} + \mathbf{K}$. Multiplying the last expression by $\mathbf{I}$ and recalling that $\Omega = I$ in the ground state, we have:

$$I(I + 1) = I^2 + \mathbf{K} \cdot \mathbf{I}.$$

Here we have made use of the fact that $\mathbf{n} \cdot \mathbf{I} = I$. Whence $\mathbf{K} \cdot \mathbf{I} = I$.

Thus,

$$\boldsymbol{\mu} \cdot \mathbf{I} = \mu_0 I + \frac{Z}{A}I.$$

Finally, the expression for the magnetic moment of the nucleus is

$$\mu = \left(\mu_0 + \frac{Z}{A} \right) \frac{I}{I + 1}.$$

The projection μ_0 has an especially simple form in the case in which the odd particle is a neutron. Since the spin of the nucleon has a projection of either $+\frac{1}{2}$ or $-\frac{1}{2}$ along the nuclear axis, $\mu_0 = \pm 1.91$.

Assuming that $Z/A \cong 0.45$ for all nuclei (Z/A changes from 0.5 in light nuclei to 0.4 in heavy nuclei), for nuclei with even Z and odd A we have the formulas

$$\mu = -1.5 \frac{I}{I + 1} \text{ or } \mu = 2.4 \frac{I}{I + 1},$$

depending on the direction of the neutron spin with respect to its orbital moment.

The formulas become more complicated in nuclei which contain an even number of neutrons and an odd number of protons because of the contribution of the orbital moment of the proton. The magnetic moment of the odd nucleon is equal to the sum of the projections of the intrinsic magnetic moment of the proton $\mu_p = 2.8$ and of its orbital magnetic moment, which is numerically equal to the orbital momentum in nuclear magnetons. The sum of the projection of the orbital moment of the proton and the projection of the spin is equal to the spin of the nucleus, I. Hence, the projection of the orbital moment is either $I + \frac{1}{2}$ or $I - \frac{1}{2}$, depending on the sign of the spin projection along the nuclear axis. Whence we have:

$$\mu_0 = \pm 2.8 + (I \mp \frac{1}{2})$$

or

$$\mu_0 = 2.3 + I \text{ and } \mu_0 = -2.3 + I.$$

Finally, the total magnetic moment of the nucleus is

$$\mu = \begin{cases} 2.75 \dfrac{I}{I + 1} + \dfrac{I^2}{I + 1}, \\[2ex] -1.85 \dfrac{I}{I + 1} + \dfrac{I^2}{I + 1}. \end{cases}$$

To make a comparison of the calculated values of the magnetic moment with the experimental value we must assume that nearly all nuclei are nonspherical. We presume at the outset that the level spectrum of a nucleus is rotational, i. e., that the nucleus is nonspherical. As has already been indicated, rotational level structures have been established only in α-active heavy nuclei and in nuclei in the rare-earth region. Hence, strictly speaking, a comparison of the formulas which have been derived can be made only for these nuclei. In the region of α-active nuclei, the magnetic moments have been measured, although very roughly, for several nuclei, and the known values of the magnetic moments of these nuclei are given in Table 5.

The nuclei U^{235} and Pu^{241} have one missing neutron. For spin $7/2$ the negative magnetic moment is -1.1, which is close to the experimental value. With spin $5/2$, the magnetic moment can be either -1.1 or 1.7, also in fairly good agreement with experiment. With spin $1/2$, the Pu^{239} nucleus should have a magnetic moment -0.5. Finally, in the Np^{237} case, which has one missing proton, with spin $5/2$ the magnetic moment should be 3.1, a value which lies within the limits of experimental error.

TABLE 5

Nucleus	Spin	Magnetic moment	
		Experimental	Theoretical
$_{92}U^{235}$	$7/2(5/2)$	-0.8 ± 0.2	$-1.1\,(1.7)$
$_{93}Np^{237}$	$5/2$	6 ± 2.5	3.1
$_{94}Pu^{239}$	$1/2$	$(-)0.4 \pm 2$	-0.5
$_{94}Pu^{241}$	$5/2$	$\pm(1.4 \pm 6)$	$-1.1\,(1.7)$

The agreement is somewhat poorer in the rare-earth region. However, in this region very little is known about the nuclear spectra or rotational levels. In hafnium, the rotational nature of the spectrum is well established, and the magnetic moment (only the absolute value is known) is 0.6 for spin $1/2$ or $3/2$, and is close to the theoretical value 0.5 or 0.9.

It is apparent that further studies of the spectra and magnetic moments of the heavy nuclei will be required before more detailed conclusions can be reached as to the applicability of a model based on a nonspherical nucleus.

An instructive example is furnished by the nucleus Bi^{209}, which is comprised of the doubly-magic nucleus Pb^{208} plus one proton. If we assume that this nucleus is spherical, its magnetic moment should be determined by the last proton. With spin $9/2$ the orbital moment of the proton should be either 4 or 5 (the $g_{9/2}$ state or the $h_{9/2}$ state). In this case the value of the magnetic moment would be 6.8 or 2.6, respectively; both values are in disagreement with the experimental value, 4.08. This situation indicates that the addition of even one nucleon to a magic nucleus is sufficient to disturb the stability of the surface.

Interesting results as to the shapes of heavy nuclei can also be obtained from a consideration of quadrupole moments. In the region with $A > 50$, the majority of nuclei have positive quadrupole moments which are several times greater than the quadrupole moment for a single particle.

We now compute the quadrupole moment of a nonspherical nucleus. The quadrupole moment of a uniformly charged system with charge Z, which has an axis of symmetry, is given by the expression

$$Q = \frac{Z}{\Omega} \int (3z^2 - x^2 - y^2 - z^2)dv,$$

where Ω is the volume of the ellipsoid.   Because of the axial symmetry

$$Q = 2Z\frac{1}{\Omega}\int (z^2 - x^2)dv.$$

In the case of a uniformly charged ellipsoid the following formula obtains:

$$Q = \frac{2Z}{5}(c^2 - a^2).$$

However, the quadrupole moment computed from this expression is not equal to the quadrupole moment of the nucleus as measured experimentally. The quadrupole moment Q_0 is the Q_{33} component of the quadrupole tensor in a coordinate system which is rigidly connected to a nucleus (in which the "3" axis is parallel to the symmetry axis of the nucleus).

Experimentally, however, one measures the average value of the component of the quadrupole tensor ($Q_{zz} \equiv Q$) of a rotating nucleus in a coordinate system which is fixed in space and usually defined by the direction of the external field. The relation between Q and Q_0 is given by the expression

$$Q = Q_0 \frac{I}{I + 1} \frac{2I - 1}{2I + 3}.$$

This expression vanishes when $I = 0$ and $I = \frac{1}{2}$.

The factor $I/(I + 1) \cdot (2I - 1)/(2I + 3)$ strongly reduces the quadrupole moment and approaches unity very slowly.   When $I = 1$ this factor is $\frac{1}{10}$ and when $I = \frac{3}{2}$ this factor is $\frac{1}{5}$; even when $I = \frac{9}{2}$ this factor is still far from unity, being $\frac{6}{11}$.

Recalling the expression for the moment of inertia

$$J = \frac{M}{5} \frac{(c^2 - a^2)^2}{c^2 + a^2},$$

we can find the semi-axis of the ellipse for every nucleus in which the quadrupole moment and rotational levels are known.

The quadrupole moments Q of the odd heavy nuclei are known from spectroscopic observations.   However, these measurements determine only the interaction energy of the nuclear quadrupole moment with the electric field due to the electron shell, and this calculation is not very reliable.   There is another more direct way of determining the magnitudes of quadrupole moments; this method is independent of the atomic properties.   If it is established that the levels exhibit rotational structure (by determining the intervals between levels), it can be shown that the lifetime for each level* is proportional to the square of the true quadrupole moment of the nucleus Q_0 (not the spectroscopic Q).   This method also makes it possible to determine the quadrupole moments in even-

---

* Also the probability for excitation by an electric field (in collisions with charged particles).

even nuclei, in which there is no spin and in which $Q = 0$.* If now, from the values of Q_0 and the moment of inertia, J, we determine the semi-axes of the ellipsoid (the ratio Q^2/J yields the sum of the squares and Q the difference) it is found that the ellipsoid is very highly elongated. However, in this analysis it is found that the volume of the nucleus is extremely small; the volume which is obtained corresponds to a value of $0.7 \cdot 10^{-13}$ cm for the constant r_0 which appears in the expression for nuclear radius $R = r_0 A^{1/3}$. On the other hand, data from electron scattering experiments give a considerably higher value for r_0: approximately $1.1 \cdot 10^{-13}$ cm. The data obtained from the spectra of mesic atoms are in agreement with this latter value. Moreover, if one computes the semi-axes of the ellipse from the radius of the nucleus and Q_0, the moment of inertia calculated from the potential flow model, turns out to be considerably smaller (three times) than the value determined from the observed spacings between rotational levels.

To some extent these discrepancies may be a result of the fact that the radius of the charge distribution in the nucleus (which determines Q) is somewhat smaller than the radius of the mass distribution (which is associated with the moment of inertia). However, the basic cause would seem to lie in the approximate character of the hydrodynamic model. This problem will require a great deal of further investigation.

Although there is no quantitative agreement, the experimental facts indicate that the heavy nuclei far from the magic nuclei are elongated ellipsoids of rotation, and that the elongation (the ratio c/a) is at least 1.5. Hence, it cannot be assumed that the field of a heavy nucleus is spherically symmetric.

APPENDIX

In this appendix we derive several formulas which have been used in the present lecture.

A. The moment of inertia for potential motion of a liquid in an axially-symmetric ellipsoidal container. We choose the z-axis as the symmetry axis of the ellipsoid and the y-axis as the axis of rotation.

Potential motion of a liquid is described by a potential φ which satisfies Laplace's equation

$$\Delta \varphi = 0.$$

The velocity of the liquid v is given by

$$v = \text{grad } \varphi.$$

The boundary conditions are that the normal component of the surface velocity of the liquid must be equal to the velocity of the container.

---

* The quadrupole moment has been measured by both methods in the odd nucleus Ta^{181}. In this case the spectroscopic value $Q = 7$ (or $Q_0 = 15$, since the spin of Ta^{181} is $7/2$) has been found to be approximately twice as great as Q_0 measured from the lifetime. Comparing these values with Q_0 for neighboring nuclei, it may be assumed that the lower value of Q_0 is the correct one.

We write the equation for the ellipse in normal form:

$$\frac{x^2 + y^2}{a^2} + \frac{z^2}{c^2} = 1.$$

The velocity of the surface is $\boldsymbol{\omega} \times \mathbf{r}$.

The direction cosines of the normals to the surface of the ellipse are proportional to x/a^2, y/a^2, and z/c^2, respectively.

Multiplying the components of the difference

$$\mathbf{v}_{sur} - \boldsymbol{\omega} \times \mathbf{r} = (\operatorname{grad} \varphi)_{sur} - \boldsymbol{\omega} \times \mathbf{r}$$

by the direction cosines of the normals and equating the expression which is obtained to zero (this is obviously the condition that the normal components vanish), we have:

$$\frac{x}{a^2}\left(\frac{\partial \varphi}{\partial x} - \omega z\right) + \frac{y}{a^2}\frac{\partial \varphi}{\partial y} + \frac{z}{c^2}\left(\frac{\partial \varphi}{\partial z} + \omega x\right) = 0.$$

This relation must be satisfied over the entire surface. We can satisfy this relation by choosing as a solution of the Laplace equation the function

$$\varphi = Axz,$$

where A is a constant. Substituting this function in the boundary conditions we have:

$$\frac{1}{a^2}(A - \omega) + \frac{1}{c^2}(A + \omega) = 0.$$

From this, having determined A, we find the potential

$$\varphi = \omega \frac{c^2 - a^2}{c^2 + a^2} xz.$$

We now compute the kinetic energy of the liquid. Setting the density equal to unity, we write:

$$T = \frac{1}{2}\int \mathbf{v}^2 dv = \frac{1}{2}\int (\operatorname{grad} \varphi)^2 dv = \frac{\omega^2}{2}\left(\frac{c^2 - a^2}{c^2 + a^2}\right)^2 \int (z^2 + x^2)dv.$$

Making the usual substitution of variables $x = a\xi$, $y = a\eta$ and $z = c\zeta$, after some simple transformations, we have

$$T = \frac{\omega^2}{2}\left(\frac{c^2 - a^2}{c^2 + a^2}\right)^2 a^2 c(c^2 + a^2)\int \xi^2 d\xi d\eta d\zeta,$$

where the integral is taken over a sphere of radius unity. Finally:

$$T = \frac{\omega^2}{2}\frac{M}{5}\frac{(c^2 - a^2)^2}{c^2 + a^2},$$

where M is the mass of the ellipsoid (equal numerically to its volume). It is apparent that the coefficient of $\omega^2/2$ is the moment of inertia in question:

$$J = \frac{M}{5} \frac{(c^2 - a^2)^2}{c^2 + a^2}.$$

B. Quadrupole moment of a uniformly charged rotating ellipsoid. To determine the factor by which we must multiply Q_0, the quadrupole moment of the fixed nucleus, to find the quadrupole moment of the rotating nucleus, Q, we form the quadrupole tensor from quantities which characterize the fixed nucleus and then from quantities which characterize the rotating nucleus.

We use the symbol n_i to denote components of the unit vector directed along the "3" axis of the fixed nucleus. The quadrupole tensor can be written

$$Q_{ik} = {}^3\!/_2 Q_0 (n_i n_k - {}^1\!/_3 \delta_{ik}).$$

The coefficient ${}^3\!/_2$ is chosen to make $Q_0 = Q_{33}$.

For the rotating nucleus Q_{ik} is expressed in terms of the spin vector $\mathbf{I}$. As in Lecture Four, we can write:

$$Q_{ik} = \frac{Q_0}{{}^2\!/_3 I(2I-1)} \left\{ I_i I_k + I_k I_i - {}^2\!/_3 \delta_{ik} I(I+1) \right\}.$$

Setting this expression equal to the mean value of the quadrupole moment computed by the preceding formula, we have

$$\overline{Q_0(n_i n_k - {}^1\!/_3 \delta_{ik})} = \frac{Q}{I(2I-1)} \left\{ I_i I_k + I_k I_i - {}^2\!/_3 \delta_{ik} I(I+1) \right\}.$$

We multiply both parts twice by the spin vector and make use of the relation $I_i n_i = \Omega$, where Ω is the projection of the spin on the symmetry axis of the nucleus (equal to the spin I), and the relations

$$\sum_{i,\,k} I_i I_k I_k I_i = [I(I+1)]^2, \quad \sum_{i,\,k} I_i I_k I_i I_k = [I(I+1)]^2 - I(I+1);$$

the latter relation follows from the commutation rules for the components of the spin vector

$$I_x I_y - I_y I_x = 2iI_z, \text{ etc.}$$

As a result we find:

$$Q_0 \left\{ \Omega^2 - \frac{1}{3} I(I+1) \right\} = \frac{Q}{I(2I-1)} \left\{ \frac{4}{3} I^2(I+1)^2 - I(I+1) \right\} = \frac{1}{3} Q(I+1)(2I+3),$$

whence

$$Q = \frac{3\Omega^2 - I(I+1)}{(I+1)(2I+3)} Q_0.$$

With $\Omega = I$ we find the formula in question:

$$Q = \frac{I(2I-1)}{(I+1)(2I+3)} Q_0.$$

Nuclear Reactions

(STATISTICAL THEORY)

The concept of individual nucleon levels is meaningful only as long as the interaction between nucleons is smaller than the distance between levels (with the same moment and parity) corresponding to different nucleon configurations. When the excitation energy increases, the density of levels grows sharply and the division of nucleons into shells loses its significance. This situation arises when we deal with heavy nuclei and relatively high excitations.

In these cases we must depart from our description based on individual-particle mechanics and go over to the other limiting case, that is to say, we must use an analysis based on statistical mechanics and thermodynamics. In doing so, however, we must keep in mind the fact that since statistical laws are valid only in systems characterized by a large number of degrees of freedom, the application of these laws to nuclei in which the number of particles is not very large obviously cannot yield accurate results.

From the point of view of statistical mechanics an excited nucleus is characterized, first of all, by an energy E, which may conveniently be measured from the ground state of the nucleus. On this scale E is simply the excitation energy. Since there are many levels in a system with a large number of degrees of freedom, the distribution of levels may be described conveniently by the mean spacing between levels at a given excitation energy $D(E)$; similarly, this distribution can be given in terms of the density of levels at a given point in the spectrum $\omega(E) = 1/D(E)$. The level density $\omega(E)$ changes with the excitation energy. As is well known, the level density increases with excitation energy.

The problem of a statistical analysis, then, is to establish a relation between the level density and the excitation energy. The relation between these two quantities is established by introducing the idea of entropy.

In statistical mechanics one establishes a relation between the entropy S and N, the number of states having approximately the same energy in a given system. This relation is the well-known Boltzmann formula

$$S(E) = \ln N$$

(in our analysis the entropy S is measured in dimensionless units).

The inverse relation then follows: the number of levels lying in a certain energy interval ΔE is given by the expression

$$N = \exp [S(E)].$$

The energy interval ΔE is determined by the magnitude of the energy fluctuations in the thermodynamic system at constant temperature. This quantity determines the inherent uncertainty in the energy of the system and can be computed for any thermodynamic system.

From the last expression we can also determine the mean spacing between levels $D(E)$.

Assuming $\Delta E = ND$, we have:

$$D(E) = \Delta E \exp [-S(E)] = \exp [-S(E) + \ln \Delta E].$$

Having determined the entropy of the system, without any further assumptions we can introduce the temperature of the nucleus, using the well-known thermodynamic expression

$$T = \left(\frac{\partial S}{\partial E}\right)^{-1}.$$

To apply statistical considerations to the nucleus and to compute the entropy of a system of nucleons we must, first of all, know the way in which the excitation energy of the nucleus depends on temperature or, what is the same thing, the heat capacity of the nucleus.

The system of nucleons which comprise the nucleus may be considered in terms of an ideal gas of particles which obey Fermi statistics.

At first glance this assumption seems strange since there is a strong interaction between nucleons. However, it is well known that a system of interacting Fermi particles has an energy spectrum very similar to the spectrum of an ideal degenerate Fermi gas. A convincing example is the spectrum of electrons in a metal. It is well known that the heat capacity of metals at low temperatures (when the contribution of the lattice to the heat capacity is negligibly small) is proportional to the temperature, just as in an ideal gas. At the same time, it is obvious that the interaction of electrons in a metal is not small. Hence, there is a basis for assuming that the heat capacity obeys a similar relation in the nuclear case.

We shall assume that the heat capacity of the nucleus depends on temperature in the same way as does the heat capacity of a degenerate Fermi gas[*]

$$c = \frac{a}{2} T,$$

where a is a constant. Since

$$c = \frac{dE}{dT},$$

---

[*] The condition for "strong degeneracy" of a Fermi gas is given by the inequality $T \ll (\hbar^2/M)\rho^{2/3}$, where ρ is the density of particles. In a nucleus $\rho = (4/3 \pi r_0^3)^{-1}$ ($r_0 \approx 1.2 \cdot 10^{-13}$ cm), whence the degeneracy condition assumes the form $T \ll 10$ Mev; it may be assumed that the temperature of the nucleus is low enough since $T \sim 1$ Mev corresponds to an excitation energy of approximately 10 Mev in a heavy nucleus.

we can determine the energy:

$$E = \frac{a}{4} T^2.$$

The entropy is determined by the relation

$$T \frac{\partial S}{\partial T} = c,$$

which yields

$$S = \frac{a}{2} T = (aE)^{1/2}.$$

The constant a should be proportional to the number of particles in the nucleus because at a given temperature the energy of the nucleus must obviously be proportional to the amount of matter. Hence,

$$a \simeq KA \text{ and } E = \frac{K}{4} A T^2,$$

where the coefficient K is relatively independent of the properties of the nucleus.

Knowing the entropy of the nucleus, we can also compute the mean spacing between levels. However, for this purpose we must calculate the entropy in a more exact fashion than is possible using the formulas given above. In this calculation logarithmic terms appear in the entropy; in the e^S expression these terms appear in the factor which multiplies the exponential. Thus, if we use the expression $S = \sqrt{aE}$ for the entropy it is impossible to calculate the factor in front of the exponential by ordinary fluctuation theory.

Since a rigorous calculation of the entropy of the nucleus has not been carried out up to this time, the quantity ΔE in the formula

$$D = \Delta E \exp [(aE)^{-1/2}]$$

cannot be computed from theory. On the other hand, the present experimental data on excited levels in heavy nuclei are inadequate for carrying out a detailed comparison of these formulas with experiment.

In discussing the experimental data on nuclear levels, we must not forget the manner in which these data are obtained. A significant number of levels are determined from experiments on slow-neutron capture. In reactions of this kind we cannot excite arbitrary levels since a slow neutron can penetrate the nucleus only when its orbital moment is zero. For example, if the initial nucleus has spin $I = 0$, the intermediate nucleus can have only one spin value, namely $1/2$. Hence, out of the large number of levels which are available in a heavy nucleus, the slow-neutron capture reaction selects only levels with lowest spin. The mean level density determined by experiments of this kind is smaller than the true density. In other experiments, for example, in experiments on inelastic scattering of fast neutrons, there are virtually no limitations due to spin; the density found in these cases is closer to the correct density.

In this connection it is interesting to examine the level density associated with a given spin value.

In accordance with the laws of statistical mechanics the probability that a system (in the present case, a nucleus) will be found in a state with a moment lying in the interval $dM_x dM_y dM_z$ about the value M is proportional to

$$dW(M_x, M_y, M_z) \sim e^{-(1/2I_0 T)(M_x{}^2 + M_y{}^2 + M_z{}^2)} dM_x dM_y dM_z.$$

The proportionality factor is determined by the normalization. Thus, for the density of levels we have

$$d\omega(M) = \omega(2\pi I_0 T)^{-3/2} e^{-(M^2/2I_0 T)} 4\pi M^2 dM,$$

where ω is the total density of levels. In quantum mechanics the moment is quantized and this formula becomes

$$\omega_I(E) = \omega(E) \frac{(2I + 1)^2}{\sqrt{\pi^3}} \left(\frac{\hbar^2}{2I_0}\right)^{3/2} \cdot \frac{1}{T^{3/2}} e^{-(\hbar^2/2I_0 T)I(I + 1)},$$

where $\omega_I(E)$ is the density of levels with a given value of moment I.

In discussing nuclear levels we have assumed above that these levels are independent. The formulas given determine only the mean distance between levels; the distribution function—that is, the probability of a given spacing between levels for a given mean spacing—has not as yet been investigated.

There is no basis for assuming that nuclear levels with different spins are in any way related; the spacing between levels with different spins should be distributed according to a random law.

However, this conclusion does not apply to levels with the same spin. The problem of the distribution for these levels has not been solved. At this point it is reasonable to assume that, in any case, levels with the same spin are not distributed according to a random law. From very general considerations one expects that levels with the same spin are distributed in such a way that the probability of a very small spacing will be very small—the levels "repel" each other. This problem requires further investigation.

Statistical considerations prove to be extremely useful for studying both nuclear reactions and nuclear decay. We start with the simplest question—the probability of α-decay. Until recently contradictory theoretical expressions have appeared in the literature. The probability for alpha decay, as is known, is determined basically by the probability that an α-particle will penetrate the potential barrier. In the theory of this effect the penetrability of the barrier (the ratio of the square of the modulus of the wave function on both sides of the barrier) is given by the expression

$$P \sim e^{-\Gamma},$$

where Γ is a quantity which depends on the energy of the α-particle and the charge of the nucleus. The decay constant λ (the reciprocal of the probability) should be

$$\lambda = \nu e^{-\Gamma},$$

where ν and λ have the dimensions of frequency. It is the quantity ν which is the source of misunderstanding. In the first papers by Gamow the nucleus was described as a potential box and the quantity ν was interpreted as the frequency with which the α-particle kept "banging" against the edge of the box. Whence the following value was obtained:

$$\nu \sim \frac{v}{R},$$

where R is a typical dimension of a nucleus and v is the "velocity" of the α-particle in the nucleus. This frequency corresponds to an energy $h\nu \approx hv/R$, which is of the order of tens of millions of electron volts. It is clear that this description is not an accurate one.

Later, another point of view was suggested; this model led to a factor which corresponded to an energy which was too small—of the order of several electron volts.

A statistical analysis of the nucleus makes it possible to obtain a rigorous formula for the factor in question. This formula is of the form:

$$\lambda = \frac{D}{2\pi\hbar} e^{-\Gamma}.$$

The coefficient $D/2\pi\hbar$ has a simple physical meaning: it is the frequency of a linear oscillator which has the same spacing between levels (equal to $h\nu$) as the nucleus in the excitation interval being considered. This quantity is of the order of several tens of kiloelectron volts.

In this form the formula relates the probability of α-decay to the total level density for the nucleus and not to the behavior of the α-particle inside the nucleus, as is the case in the Gamow formula. Hence, in deriving the formula there is no need for discussing the existence of the α-particle inside the nucleus.

The thermodynamic approach proves to be especially useful in analyzing a second type of nuclear reaction—the emission of particles from an intermediate nucleus. An excited intermediate nucleus can be considered a heated body from which particles are evaporated. The evaporation probability can be calculated in exactly the same way as the evaporation rate for ordinary liquids.

In this picture we consider the nucleus to be in thermodynamic equilibrium with the gas of evaporated particles. In a system of this kind the number of evaporated neutrons is equal to the number of absorbed neutrons; moreover, the number of absorbed neutrons can be calculated easily. Since we are not interested in the absolute probability, but only in the relative probability of emission of neutrons with different energies (neutron evaporation spectrum), we neglect all factors which do not depend on energy. The energy distribution of the neutrons in the gas which surrounds the nucleus is given by

$$f(\mathcal{E})d\mathcal{E} \sim e^{-(\mathcal{E}/T)}\mathcal{E}^{1/2}d\mathcal{E}.$$

The number of neutrons with energy $\mathcal{E}$ which collide with the surface of the nucleus in unit time is proportional to the number of neutrons $f(\mathcal{E})$ and the neutron velocity (i. e., $\mathcal{E}^{1/2}$). Thus, the number of neutrons which strike the nucleus and have energy $\mathcal{E}$ is given by

$$\nu(\mathcal{E})d\mathcal{E} \sim e^{-(\mathcal{E}/T)}\mathcal{E}d\mathcal{E}.$$

However, not every collision of a neutron with a nucleus results in the formation of an intermediate nucleus. We shall denote the probability for an event of this kind by $\xi(\mathcal{E})$. The coefficient ξ is called the *sticking coefficient* (in molecular physics the analogous quantity is called the accommodation coefficient). Collecting all factors and using the fact that the number of absorbed neutrons is equal to the number of emitted neutrons, we can write the distribution relation for evaporated neutrons in the form

$$n(\mathcal{E})d\mathcal{E} \sim \xi(\mathcal{E})\mathcal{E}e^{-(\mathcal{E}/T)}d\mathcal{E}.$$

At energies of several million electron volts the sticking coefficient ξ approaches unity. If we assume $\xi = 1$, the mean energy of the evaporated neutrons becomes:

$$\mathcal{E}_{\text{mean}} = 2T.$$

At low energies the coefficient ξ is considerably different from unity; at energies close to zero (thermal region), the sticking coefficient is proportional to velocity:

$$\xi \sim \mathcal{E}^{1/2} \qquad (\mathcal{E} \rightarrow 0)$$

whence

$$\mathcal{E}_{\text{mean}} \simeq {}^5/_2 T \qquad (\mathcal{E} \rightarrow 0).$$

Thus, the mean energy of the evaporated neutrons satisfies the relation

$$2T < \mathcal{E}_{\text{mean}} < 2.5T.$$

It is important that the energy of the evaporated particles be considerably smaller than the excitation energy. Under these conditions the excitation energy is dissipated in the emission of several nucleons with comparatively small energy rather than in the emission of a single fast nucleon.

The evaporation of neutrons has one distinguishing feature by which it differs from evaporation in ordinary liquids. The energy of a molecule which is evaporated from the surface of a liquid is small compared with the energy of the total liquid, so that the temperature of the liquid remains constant (isothermal evaporation). In the case of nuclear evaporation, however, the neutron carries off (considering its binding energy) a considerable fraction of the available energy. Hence, to obtain really accurate formulas we must indicate the proper nuclear temperature. It is easy to show (inasmuch as the derivation of the formula was based on an analysis of the inverse problem—the equilibrium of the final nucleus with the neutron gas) that the quantity T which appears in the formulas is the temperature of the final nucleus.

The evaporation of charged particles—protons, deuterons and heavier nuclides—can be considered by an analysis similar to that used in the evaporation of neutrons. Proton evaporation differs from neutron evaporation because the Coulomb repulsion of the protons means that the absorption probability $\xi(\varepsilon)$ is smaller at lower energies and that the limiting value $\xi = 1$ is reached at higher energies.

From this point of view the case of the deuteron is of special interest. It is known that there is a comparatively large number of deuterons in cosmic rays. These are undoubtedly the result of nuclear reactions. It seems strange, at first glance, that a nucleus can emit a comparatively unstable particle such as the deuteron in one piece.

A consideration of deuteron evaporation from the nucleus shows that the probability for evaporation is determined only by the excitation energy (temperature) of the final nucleus and is, consequently, virtually independent of the binding energy of the deuteron. Even if the binding energy of the deuteron were zero its evaporation probability would not vanish. These same considerations would seem to explain the fact that the Be^8 nucleus is emitted from nuclei in nuclear reactions; this nuclide is extremely unstable and decays into two α-particles in a time of approximately 10^{-14} sec (decay energy approximately 50 kev).

Nuclear Reactions

(OPTICAL MODEL. DEUTERON REACTIONS)

In this lecture we shall consider nuclear reactions from an entirely different point of view. To eliminate the complications which arise because of the Coulomb field of the nucleus we shall consider neutron reactions only.

Consider a neutron which collides with a heavy nucleus. There is considerable probability that the neutron will pass through the nucleus without losing energy, suffering only a change in the direction of its motion. This process corresponds to elastic scattering. However, a neutron can also be absorbed, thereby producing an intermediate nucleus which can then decay by various modes. In particular, the intermediate nucleus can emit a neutron with smaller energy (inelastic scattering) or with energy equal to that of the initial neutron (re-emission of a neutron).

If we are not concerned with the intermediate nucleus, all events which occur in the collision of a neutron with a nucleus can be divided into two types—scattering (without the formation of an intermediate nucleus) and absorption (formation of an intermediate nucleus).

Using this picture we can consider the collision of a neutron with a nucleus in terms of the passage of the neutron through a certain medium—nuclear matter—and consider this entire problem by analogy with the passage of light through a medium which exhibits refraction and absorption.

Just as in the case of an optical medium, we can introduce an index of refraction, which is a complex quantity if there is absorption; thus, we require two quantities to characterize the "optical" properties of nuclear matter (in the following we will speak of the optical properties of the nucleus). This approach to nuclear reactions is called the *optical model of the nucleus*.

In order to determine these parameters we turn to the Schrödinger equation for the neutron. According to quantum mechanics the motion of a nucleon is described by a wave function which is associated with a certain potential. If the potential is real we are dealing with neutron scattering; on the other hand, as we shall now see, neutron absorption is to be associated with a complex potential.

To verify this statement we introduce the expression for a flux of particles, using a wave equation with complex potential. We will proceed exactly as in deriving the expression for a flux of particles using an ordinary potential.

We write the wave equation for a particle in a field described by a complex potential $U - iV$ (U and V are real functions):

$$E\psi = -\frac{\hbar^2}{2m} \Delta\psi + U\psi - iV\psi,$$

where E is the energy of the particle. We now write the conjugate equation:

$$E\psi^* = -\frac{\hbar^2}{2m} \Delta\psi^* + U\psi^* + iV\psi^*.$$

Multiplying the first equation by ψ^* and the second by ψ, we take the difference of these two expressions. After some elementary manipulation we have

$$-\frac{\hbar^2}{2m} \text{div} (\psi^* \text{ grad } \psi - \psi \text{ grad } \psi^*) - 2V\psi\psi^* = 0.$$

Since $\psi\psi^*$ is the density of particles ρ, and $\hbar/2mi(\psi^* \text{ grad } \psi - \psi \text{ grad } \psi^*)$ is the current density j, the last relation can be written:

$$\text{div } j = -2\rho V.$$

The left-hand side of this expression is the divergence of a current; from the physical meaning of this quantity, we see that it vanishes if particles are neither produced nor absorbed. Hence, the right-hand part of the equation is nothing else than the source intensity. Thus, the absorption of particles per unit time per unit volume is determined by the quantity $2V\rho = 2V\psi^*\psi$.

The imaginary part of the complex potential is then the absorption coefficient.

In order to formulate a theory for the passage of a neutron through the nucleus starting from these ideas, we must have some information for determining both parts of the potential (real and imaginary). There is no reason to assume that these quantities are independent of energy; on the contrary, they are undoubtedly functions of the neutron energy, and must be determined from experiment.

To make a clear distinction between processes connected with the formation of an intermediate nucleus and those connected with true elastic scattering we must first make sure that the probability for the re-emission of a neutron with the initial energy is small. In turn, this requirement implies that the intermediate nucleus must be capable of decay by many modes. This condition is satisfied if the neutron has an energy of the order of several million electron volts; under these conditions there are a number of possible reaction channels and excited states available for the final nucleus.

Transforming the Schrödinger equation as written above, we can show that in the optical model the neutron motion is given by the equation

$$\Delta\psi + \frac{2m}{\hbar^2} (E - U + iV)\psi = 0.$$

This equation is analogous to the wave equation in vacuum

$$\Delta\psi + k^2\psi = 0,$$

where **k** is the propagation vector.

Formally, we can say that the magnitude of the propagation vector, which is

$$k = (2mE)^{1/2}/\hbar$$

in vacuum, becomes a complex quantity in the nucleus:

$$K = K_1 + iK_2 = \frac{(2m)^{1/2}}{\hbar}(E - U + iV)^{1/2}.$$

By analogy with optics, we can speak of a complex index of refraction for nuclear matter:

$$n = \frac{1}{k}(K_1 + iK_2).$$

This quantity, however, is usually not employed in nuclear physics.

We now examine the physical meaning of the complex propagation vector. When the neutron is inside the nucleus the particular solution of the wave equation can be written in the same form as the vacuum solution, i. e., a plane wave (but with a complex propagation vector).

Limiting ourselves to the one-dimensional problem, we can write

$$\psi = e^{iKk} = e^{iK_1x}e^{-K_2x}.$$

It is obvious from this expression that the particle absorption depends on K_2.

In order to relate the complex propagation vector to the quantities which are determined from neutron scattering experiments, we must solve the problem of diffraction of a neutron wave on a nucleus with given optical properties and radius.

We must determine three quantities from experiment at each value of the energy: the real and the imaginary parts of the potential energy, and the radius of the nucleus (independent of energy); thus, we require rather detailed experiments on both elastic and inelastic scattering to determine all quantities if additional assumptions are to be avoided.

It should be noted that in general these constants can be different for protons and neutrons. This difference may be due, first, to Coulomb forces, and second, to the fact that the number of protons and neutrons is different in heavy nuclei. At the present time this very interesting question has not been investigated experimentally.

In order to solve the problem of neutron diffraction on the nucleus we must find the solution of the wave equation inside and outside the nucleus.

The inside and outside solutions must be matched at the boundary. The derivatives must also be matched at the boundary. Usually the solution is expanded in a series of spherical functions and the matching conditions yield algebraic equations for the coefficients of the series. The solution of this problem requires the calculation of a comparatively large number of terms and proves to be rather difficult.

The problem can be simplified considerably if we consider high energy neutrons; in this case the wavelength of the neutron is small compared with the dimensions of the nucleus. The neutron wavelength λ_n is given in centimeters by the formula

$$\lambda_n = \frac{4.5}{\sqrt{E}} 10^{-13},$$

where E is given in millions of electron volts. The radius of the nucleus R is of the order of $5 \cdot 10^{-13}$ cm; hence, the condition $\lambda_n \ll R$ is satisfied at neutron energies of several hundreds of millions of electron volts. Using neutrons of this kind, we can solve the problem by a method which is very similar to that used in optics to solve Fraunhofer diffraction, for example, the diffraction of a parallel beam of light at a circular aperture.

For simplicity we shall consider a spherical nucleus. If the nucleus is not spherical, the problem is considerably more difficult. Suppose that the neutron beam is incident along the x-axis. The incident neutron beam is described by the plane wave e^{ikx}, where k is the magnitude of the neutron propagation vector in vacuum. The amplitude of the wave function is taken as unity.

Consider a plane, perpendicular to the incident neutron beam and located directly behind the nucleus (i. e., on the side away from the incident beam). We will call this plane the "shadow" of the nucleus, i. e., that region which neutrons moving along rectilinear trajectories parallel to the propagation vector $\mathbf{k}$ can reach only by passing through the nucleus. As an approximation we assume that at all points of the plane outside of the shadow region the wave function is a constant whose value is that which obtains in the absence of the nucleus.

Since we can always multiply the wave function by an arbitrary phase factor we take the value of the phase to be zero at all points lying outside the shadow region. This means that we have chosen the z-coordinate in such a way that the plane corresponds to $z = 0$. Thus the wave function

$$\psi = e^{ikz}$$

becomes unity outside the shadow region.

To compute the wave function in the shadow region we note that inside the nucleus (over a path of length Δ, which is different for different parts of the shadow), the propagation vector is not k, but $K_1 + iK_2$. If this segment of path were in vacuum the phase would be changed by an amount $k\Delta$. Inside the nucleus, however, the actual phase change is $(K_1 + iK_2)\Delta$. Thus, in passing through the nucleus the neutron wave function undergoes an additional phase shift, equal in magnitude to

$$\delta = (K_1 + iK_2 - k)\Delta.$$

We have assumed that the wave function is unity outside the shadow; in the shadow the wave function is then

$$\psi_{\text{shadow}} = e^{i(K_1 - k)\Delta} e^{-K_2\Delta}.$$

From a knowledge of the wave function we can also compute the cross sections. It is simplest to use the absorption cross section. The neutron density (equal to $\psi^*\psi$) in various points of the plane is simply

$$\psi\psi^* = \begin{cases} e^{-2K_2\Delta} \text{ in the shadow region,} \\ 1 \text{ at other points} \end{cases}$$

If $K_2 = 0$, the density is unity at all points in the plane.

It is apparent that the difference between unity and $e^{-2K_2\Delta}$ corresponds to the absorption of a neutron by the nucleus. The attenuation of the beam which takes place as a result of absorption by the nucleus is thus given by $1 - e^{-2K_2\Delta}$. From this we easily obtain the cross section. Specifically, integrating the attenuation function over the cross section of the beam, i. e., over the plane, we have

$$\delta_{\text{abs}} = \int (1 - e^{-2K_2\Delta})dxdy,$$

where the integration is actually carried out over the shadow region; obviously Δ depends on the coordinates x and y.

It is somewhat more difficult to obtain the formula for elastic scattering. In the absence of a nucleus the neutron wave function would be a plane wave and, in a coordinate system in which the z-axis is parallel to the wave vector, the wave function would be independent of the x and y coordinates (the wave function is simply e^{ikz}). The presence of the nucleus changes the wave function in such a way that it becomes a function of the coordinates of the plane. Thus, the wave function is no longer a plane wave, i. e., it no longer describes a parallel beam of particles. Inasmuch as the neutron energy is not affected (processes which lead to a change of energy are taken into account by the absorption), the new wave function describes particles which are scattered in different directions. In order to distinguish the different particles by direction, in accordance with the laws of quantum mechanics, we must expand the wave function in the xy plane in a Fourier integral. The terms in this expansion (more precisely, the squares of the moduli) then determine the particle flux in various directions. The Fourier expansion must be carried out in the two variables k_x and k_y, corresponding to the two coordinates of the xy plane. If there were no scattering this expansion would contain only one component, since $k_x = k_y = 0$.

Thus we write the wave function in the form

$$\psi = (2\pi)^{-1} \int a(k_x, k_y)e^{i(k_xx + k_yy)}dk_xdk_y.$$

Then, in accordance with the laws of quantum mechanics, $|a(k_x, k_y)|^2$ is the quantity which determines the fraction of neutrons with propagation-vector projections on the x-axis and y-axis in the intervals k_x, $k_x + dk_x$ and k_y, $k_y + dk_y$. From the Fourier theorem

$$a(k_x, k_y) = (2\pi)^{-1} \int \psi(x, y)e^{-i(k_xx + k_yy)}dxdy.$$

The function $\psi(x, y)$, which must be substituted in this formula as has been mentioned above, is $\exp [i(K_1 - k)\Delta - K_2\Delta]$ in the shadow region, and unity at other points. The scattering cross section is given by the expression

$$d\sigma(k_x,k_y) = (2\pi)^{-2}\left| \int \psi(x,y) \exp [-i(k_x x + k_y y)]dxdy \right|^2 dk_x dk_y.$$

To transform from the distribution over propagation-vector components (k_x, k_y) to a distribution over scattering angle, we introduce the vector $\boldsymbol{\theta}$ in the xy plane; the magnitude of this vector determines the scattering angle. Thus, $k_x = k\theta_x$ and $k_y = k\theta_y$. Since the scattering is independent of the azimuthal angle, the cross section depends only on the length of the vector $\boldsymbol{\theta}$. The integral which determines the scattering cross section can be computed, although the expression which is obtained is extremely complicated.

To obtain the total scattering cross section we can integrate the differential cross section over angle or propagation-vector components k_x and k_y. However, there is a simpler method.

We have seen that the wave function ψ differs from unity only over a small section of the plane. Thus we can write ψ in the form

$$\psi = 1 + (\psi - 1).$$

Written this way, the unity term can be interpreted as the incident wave propagating along the axis (we may recall that we are speaking of the value of the wave function at the given plane). Then the difference $\psi - 1$ describes the change in the wave function due to diffraction at the nucleus. The above expression can be interpreted as the sum of the incident and scattered waves. The total number of elastically scattered neutrons can be obtained if we integrate $|\psi - 1|^2$ over the plane. Thus we have

$$\sigma_{\text{elastic}} = \int |\psi - 1|^2 dxdy.$$

As in the case of absorption, the integration in this formula is actually carried out only in the shadow region. Since $|\psi| < 1$ for absorption, any absorption must be accompanied by elastic scattering.

The total cross section for the interaction of neutrons with the nucleus is written as the sum

$$\sigma_{\text{total}} = \sigma_{\text{elastic}} + \sigma_{\text{abs}}.$$

Combining the expressions for both cross sections, we have:

$$\sigma_{\text{total}} = \int (1 - |\psi|^2 + |\psi - 1|^2)dxdy$$

or, taking out the expressions for the moduli:

$$1 - |\psi|^2 + |\psi - 1|^2 = 2 - 2\text{Re}\psi,$$

where Re denotes the real part.

Thus,

$$\sigma_{\text{total}} = 2 \int (1 - \text{Re}\psi)dxdy.$$

We can also express σ_{total} in terms of the functions K_1 and K_2. Substituting the expression for ψ and integrating over the shadow region, we have:

$$\sigma_{\text{total}} = 2 \int (1 - e^{-2K_2\Delta} \cos 2(K_1 - K_2)\Delta)dxdy.$$

Thus, we have obtained all the relations needed to describe the interaction of a neutron with a nucleus. At each neutron energy the values of the two cross sections, for example, σ_{total} and σ_{abs}, are related to the values of K_1 and K_2 at this energy by the above formulas.

At the present time, the experimental data are not sufficient to determine all three optical characteristics of the nucleus: the complex index of refraction $k^{-1}(K_1 + iK_2)$ (which is a function of energy) and the radius of the nucleus (which is independent of energy). To determine these quantities, it is obviously not sufficient to know only the magnitudes of the total and elastic cross sections for any given nucleus; it is also necessary to know at least the elastic-scattering angular distribution.

At this point we should add that the nuclear radii determined from neutron cross sections are considerably larger than those found in electron scattering and mesic atoms. However, one must remember that this difference does not necessarily imply a discrepancy, since we are actually speaking of effects in which the radius may not be defined in the same way. Moreover, the radius obtained from neutron scattering should be larger than the radius of the nucleus by an amount equal to the radius of the neutron itself (range of nuclear forces).

Several conclusions can be reached from neutron scattering at high energies. If the nucleus were completely opaque to neutrons the absorption cross section would be πR^2. Indeed, in this case ψ would become zero in the shadow region and the integral over $1 - |\psi|^2$ would be equal to the area of the shadow. Experimentally, the absorption cross section proves to be smaller than this value. This situation means that the nucleus is a semi-opaque medium for fast neutrons.

We consider the case in which the nuclear refraction is large (this is the actual physical case), so that

$$(K_1 - k)\Delta \gg 1.$$

In the shadow region the quantity $\text{Re}\psi = e^{-K_2\Delta} \cos (K_1 - k)\Delta$, which appears under the integral sign and determines the total cross section, changes sign many times. Since this function oscillates, in the integration over the shadow region the total integral is small. Hence, in cases of large refraction $\text{Re}\psi$ becomes effectively zero, and the formula for the total cross section assumes the form

$$\sigma_{\text{total}} = 2\pi R^2.$$

The ratio of cross sections for elastic scattering and absorption obviously depends on the magnitude of the absorption.

We have seen that at large absorption the absorption cross section approaches πR^2; hence, in the limiting case

$$\sigma_{\text{abs}} = \sigma_{\text{elastic}} = \pi R^2 \quad \text{for} \quad K_2 \Delta \gg 1.$$

On the other hand, at small K_2 there is essentially no absorption, and the scattering cross section becomes the same as the total cross section:

$$\sigma_{\text{abs}} = 0,$$

$$\sigma_{\text{elastic}} = 2\pi R^2 \quad (K_2 \Delta \ll 1).$$

We have discussed only the simplest properties of the optical model. In an actual analysis of the experiments it must be considered in greater detail. First of all, the considerations given above must be extended to include the case in which the incoming particle is a charged particle. This situation leads to a certain complication in the formulas.

The greatest difficulty appears if we give up the assumption that the nucleus is a sphere with sharply defined boundaries. In an actual nucleus the density does not change sharply, but falls off more or less gradually at the boundary. Moreover, in heavy nuclei we must take account of the fact that the nucleus is not spherical. The introduction of these refinements makes the formulas complicated and makes an analysis of the experimental results very difficult.

Interesting results are obtained if one takes account of the spin of the incoming particles. Because of spin-orbit coupling (which has already been discussed in our review of the light nuclei), particles with different spin orientations are scattered differently; experimentally this leads to a polarization of the scattered particles. An analysis of this effect can also be carried out within the framework of the optical model.

There is one type of nuclear reaction which differs from the usual reaction since the nucleons involved exhibit a resonance effect. We are speaking of reactions induced by deuterons.

The deuteron is distinguished from all other nuclei by its small binding energy. The binding energy of the deuteron is 2.23 Mev. Because of its small binding energy the deuteron is an "extended" system. The mean distance between the proton and neutron in the deuteron is approximately $4 \cdot 10^{-13}$ cm. This value exceeds the range of the interaction forces between nucleons; hence the deuteron (like a nucleon) cannot be considered a point particle. This situation leads to a number of characteristic features in deuteron reactions.

Different processes can occur in the collision of a deuteron with a nucleus. In the first place a deuteron can be scattered elastically. The deuteron can also enter the nucleus as a whole, forming an intermediate nucleus. Both of these processes have no unusual features and will not be considered further.

We consider reactions of the (d, n) or (d, p) type. In these reactions the outgoing particle is part of the initial deuteron. These reactions can go via the usual channel—through an intermediate nucleus—but there is considerable probability (in many cases higher than the probability for formation of an intermediate nucleus) of another type of reaction. Because of its large size, the entire deuteron may not enter the nucleus; it is sufficient if either the proton [(d, n) reaction] or the neutron [(d, p) reaction] does so. The second process has some unique features.

In contrast to the deuteron as a whole, the neutron does not have to overcome Coulomb repulsion in order to penetrate the nucleus. Hence, at rather small deuteron energies (from several hundreds of kiloelectron volts to many millions of electron volts) the reaction being considered is more probable than the entry of the whole deuteron into the nucleus. A similar situation obtains in the (d, n) reaction. Although the Coulomb barrier for the proton is the same as for the deuteron, the proton mass is only $1/2$ that of the deuteron and the probability for a proton to penetrate the barrier is greater than that of the deuteron. Deuteron reactions which take place without the formation of an intermediate nucleus are called Oppenheimer-Phillips reactions, from the names of the authors who first suggested (although not completely correctly) the theory of this effect.

The interaction of a deuteron with a nucleus need not necessarily terminate in the capture of one of the nucleons; the interaction may simply lead to the splitting of a deuteron—the (d, np) reaction. This reaction can also take place outside the nucleus. The last reaction has been analyzed theoretically; however, it remains almost entirely uninvestigated experimentally.

At energies of several millions of electron volts the (d, p) and (d, n) reactions can be used to study characteristics of the levels of light nuclei. This possibility was first suggested by Butler* who also gave an analysis of the effect. Although the actual calculations carried out by Butler are unconvincing from a theoretical point of view and the theory of the effect has not been fully formulated, the qualitative results obtained by Butler are undoubtedly correct and are corroborated by numerous experiments. It would seem that these results are highly insensitive to the actual assumptions used in the calculations.

For example, consider the (d, p) reaction in a certain nucleus. From a knowledge of the energy of the proton after the reaction and the initial deuteron energy it is possible to determine at which level of the final nucleus the neutron is captured. If the nucleus is a light nucleus the spacing between levels is large and this identification is unique. Naturally this type of reaction can go via the intermediate nucleus; in this case, however, the protons are emitted uniformly in all directions. In the case in which only one neutron is "stuck," the protons have a sharply defined angular distribution, the shape of which can be used to obtain information on the characteristics of the nuclear levels. The process in which the intermediate nucleus divides, which occurs in parallel with the above process, results in the appearance of an isotropic background which can be distinguished in principle. The angular distribution of the emitted protons is determined by the angular momentum carried into the nucleus by the neutron. If the neutron does not carry such momentum, the proton distribution has a narrow peak in the forward direction. This is the case which can most clearly be established in experiment.

If the maximum in the angular distribution does not lie at zero angle, it means

---

* S. T. Butler, Proc. Roy. Soc. (London) **A 208**, 559 (1951).

that the neutron has transferred to the nucleus a momentum $\Delta l > 0$. The theory allows us, in a very rough way, to establish the position of the maxima for $\Delta l = 1, 2$. It is difficult to distinguish between higher values of Δl.

A large number of experiments have made it possible to obtain new (or to verify known) data on nuclear levels. There is considerable interest in the investigation of successive (from nucleus to nucleus) nuclear levels in regions where shells have not been firmly established. It is clear that the same results can also be obtained in studies of (d, n) reactions in proton capture.

The reaction which is the inverse of the one described above is also suitable for this purpose. If a proton enters a nucleus it can attract one of the neutrons in the nucleus, and a deuteron can be formed. As a result we have the (p, d) reaction. In the same way a neutron may eject a proton from the nucleus and cause the (n, d) reaction. These two reactions are called "pick up" reactions. These are the inverse reactions to the (d, p) and (d, n) reactions, and their probabilities can be expressed in terms of the probabilities for these latter reactions, using statistical relations (the principle of detailed balancing). Hence, all results which can be obtained from one reaction can also be obtained from the other.

Similar reactions take place at high deuteron energies. In this region the (d, p) and (d, n) reactions are called stripping reactions. The nature of the effect is the same. However, the theoretical analysis is simpler. At small deuteron wavelengths the nuclear target can be considered a circular, opaque (or semi-opaque) object which absorbs all particles incident upon it—neutrons or protons. The stripping effect has been investigated in detail, both theoretically and experimentally.

It is of interest to note certain features of (p, n) or (n, p) reactions in light nuclei. These reactions are the inverse of each other (for a suitable choice of target), and hence their properties are similar.

If, for example, a proton strikes a nucleus, then, as in the deuteron reaction, two reaction channels are possible. The reacton which proceeds via the intermediate nucleus will give neutrons which are distributed isotropically over angle; the other reaction is the direct ejection of a neutron from the nucleus. In this case conservation of energy requires (if the proton is not fast) that the incoming proton stick in the nucleus. In this reaction the neutron angular distribution depends on the proton state and the neutron state in the nucleus. The study of these reactions is of a special interest in nuclei in which these states are different since transitions between mirror nuclei lead to the trivial replacement of a proton by a neutron and vice versa—in this case the emitted neutron is essentially the "extension" of the proton, and the distribution exhibits a maximum at zero degrees.

Thus we see that the study of reactions which occur without the formation of an intermediate nucleus provides a unique method for studying the structure of the nucleus.

LECTURE NINE

π-Mesons

*I*n many cases the collision of a nucleon with a nucleus at high energies is characterized by the production of new particles—π-mesons. Processes involving the production of π-mesons and the interaction of these particles with nucleons and nuclei have been intensely studied in recent years with the increasing availability of huge accelerators which can accelerate protons to energies greater than 10^9 ev.

Before discussing the properties of π-mesons themselves, we shall establish certain general considerations which pertain to the production of particles and the transformation of particles from one type to another.

Any process which takes place in a quantum system must conserve the total energy, the total angular momentum and the parity of the system.

In computing the angular momentum of a system, in addition to the orbital momenta of the particles, we must take account of the inherent moments—the particle spins. In exactly the same way, in applying conservation of parity, we must consider not only the parity associated with the orbital momentum L, which is $(-1)^L$, but also the (intrinsic) parity of the particles. Obviously, the intrinsic parity of particles can be considered only in cases in which the process does not result in the creation or annihilation of particles; in these cases the "orbital" parity is conserved separately.

The intrinsic parity of particles is not uniquely defined. For example, it is possible, without violating any of the conservation laws, to change the parity for all particles in a system by multiplication by $(-1)^Z$, where Z is the charge expressed in units of electronic charge (by parity we are to understand the quantity $+1$ or -1, by which the wave function is multiplied in making a change from a right-handed coordinate system to a left-handed system). The total parity of all the particles before the reaction is multiplied by $(-1)^{\Sigma Z}$, where the sum is taken over all particles. Similarly, the total parity of all the particles after the reaction is multiplied by $(-1)^{\Sigma Z'}$. By virtue of charge conservation ΣZ (before the reaction) is equal to $\Sigma Z'$ (after the reaction); hence, this transformation does not violate the conservation law.

There is, however, a case in which the parity of the particles can be given absolutely. We are referring to neutral particles, primarily photons. The parity of the π^0-meson can also, in principle, be established from an investigation of the reaction in which it decays into two photons.

In addition to the parity associated with coordinate reflection—space parity—

we can consider the properties of a system under time reversal. In this case, in place of time reflection it is convenient to consider the simultaneous "reflection" of the space and time coordinates, rather than the time coordinates alone.

We can classify particles in terms of behavior with respect to a change of sign for all coordinates (space and time). Exactly as in the case of space parity, we formulate a conservation law for the time parity of the system: the total time parity of the particles cannot be changed in any process. Here we are speaking of the total parity of the particles; in contrast with ordinary parity, there is no time parity associated with the orbital motion and we are discussing here only the conservation of the internal or intrinsic parity of the particles.

The time parity of particles has not been studied to any great extent.

We may say that the time parity of the photon is $+1$. This statement follows from the fact that the photon can appear without the annihilation or creation of any other particles in a system, i. e., without a change in the time parity of these particles. The same considerations pertain to the π^0-meson.

It is of great interest to study the conservation law for time parity in order to understand the absence of a number of processes in nature which are formally possible. Such processes, for example, are those which lead to the annihilation of nucleons, say

$$n + p \rightarrow \pi^+ + \pi^0$$

or

$$p + e^- \rightarrow 2\gamma.$$

These processes do not occur in nature; indeed, it is only because they do not occur that we have stable nuclei.

It might be assumed that the hydrogen atom does not transform into two quanta (the last reaction) because the time parities of the electron and proton are different (similarly for the time parities of the proton and neutron). Then the hydrogen atom could not transform into an even system (2 quanta) since it, itself, would be an odd system. However, having prohibited this transformation on this basis we could not prohibit the decay of an even system—hydrogen molecule or deuterium atom—into a pair of photons. Thus, the simple explanation is unsatisfactory.

Both parities which we have been discussing pertain to charged particles and neutral particles—nothing specifically related to the charge has been introduced in any of the above considerations. In neutral particles there is one other specific characteristic, related to the symmetry properties of the quantum-mechanical equations for neutral particles.

Quantum-mechanical equations allow a unique transformation, called charge conjugation, in which the particle is replaced by the anti-particle. The anti-particle for the electron is the positron; the anti-particle for the π^+-meson is the π^--meson. In October of 1955 the anti-proton was discovered by Wiegand, Ypsilantis, Segre and Chamberlain. The anti-neutron—a neutral particle with positive magnetic moment—has not yet been observed, but it is clear that it must exist.

The property of the equations to which reference has been made is the fact that the equation remains unchanged if we replace all particles of this system by their anti-particles and change the sign of the electromagnetic field.

It is obvious when we speak of a particle—anti-particle pair, which of them we call the particle and which we call the anti-particle is really a question of notation and has no physical significance.

We must be more specific in what is meant by a neutral particle. Strictly speaking, the neutron is not completely neutral—it has a magnetic moment. Hence, it is possible to have an anti-neutron—a particle with magnetic moment of the opposite sign. On the other hand, the photon is a neutral particle in the strict sense of the word. The π^0-meson is also a neutral particle. We may define a neutral particle rigorously by saying that it is a particle *which is identical with its own anti-particle.*

According to this definition a particle may not be a true neutral particle even if it has no charge or magnetic moment. The simplest example of this type is the hydrogen atom in a state in which the total spin (electron—nucleus) is zero. Although the hydrogen atom in this state has neither total charge nor moment, it is still not identical with its own anti-particle—an anti-proton with a positron in the K-orbit. On the other hand, the photon has no anti-particle—it is identical with its own anti-particle and is a neutral particle in the true sense of the word.

By this definition of a neutral particle we imply another property of neutral particles—charge parity. In making a transformation to the charge-conjugate system neutral particles are transformed into themselves. Hence (if we limit ourselves to particles with integral spin) the wave function for neutral particles may either change sign or not change sign. Thus two types of neutral particles arise—charge-odd and charge-even.

The notion of charge parity has direct practical application. It is easily seen that the photon is a charge-odd particle. In making the transition to the charge-conjugate equation the field component changes sign. But the field components are the quantities which are associated with photons in quantum mechanics. Consequently, the wave function of the photon is characterized by sign as far as charge conjugation is concerned, and thus the photon is charge-odd.

On the other hand, as will be seen below, the π^0-meson is charge-even. We may also note that a system of two photons or two mesons (or any two identical neutral particles) is always charge-even.

Up to this point we have discussed only particles with integral spin. The case in which a particle has half-integral spin exhibits certain peculiarities. Particles of this type are described by special mathematical quantities called *spinors.* Whereas scalars, vectors and tensors come back to their original values when the coordinate system is rotated through 360°, a spinor reverses sign in such a transformation. This property of spinors does not lead to any inconsistency since the wave function itself is not measured in experiment; however, this property turns out to have an important effect on the reflection of wave functions.

For example, consider reflection of the spatial coordinates. Suppose that under a transformation of this kind the wave function is changed in accordance with some rule which we will write symbolically:*

$$\psi \rightarrow I\psi.$$

We repeat this transformation and return to the original coordinate system; the total operation can be written in the form:

$$\psi \rightarrow I(I\psi) = I^2\psi.$$

If we are dealing with a scalar, vector or tensor, we require that $I^2 = 1$, since the wave function must return to its original value. Whence it follows that two types of particles are possible, corresponding to the two signs for the root of unity.

In the case of particles with half-integral spin the wave function, as has already been indicated, may either return to the initial value or may change sign. Correspondingly, there are two possible cases: either we can consider the double inversion as the identity operation, in which case $I^2 = 1$, or we can consider the case in which the wave function changes sign under double inversion, in which case $I^2 = -1$. Thus, in the case of spinors we have the possibility of four different inversion transformations and, correspondingly, the possibility of four "parities." However, it is obvious that all four of these possibilities cannot be realized simultaneously. The square of the inversion operator must have a single definite value for all spinors; this, however, cannot be determined theoretically. Each of the two possible values $I^2 = \pm 1$ is consistent with the formalism of the theory. The determination of this fundamental property of the inversion operator can be made only on the basis of experiment.

There is, however, an unexpected possibility for determining the value of I^2 from experiment. It turns out that if the square of the inversion operator $I^2 = 1$ the wave functions of the particle and anti-particle transform differently under inversion; if, however, $I^2 = -1$ the transformation laws are the same. This situation implies that a true neutral particle with half-integral spin can exist only if the second relation holds true. If the first law applies, obviously the particle and anti-particle cannot be identical since the transformation relations for their wave functions are different.

Thus, to decide definitely which type of transformation to assign to the wave functions we must determine whether or not a true neutral particle exists in nature. A true neutral particle of this type could be the neutrino. The neutrino, by definition, has no charge; there is also a high probability that it has no magnetic moment.

According to the usual β-decay description, the neutrino (ν) appears in the emission of a positron from the nucleus, while the anti-particle—the anti-neutrino ($\bar{\nu}$) appears in the emission of an electron. If the neutrino is a true neutral particle the neutrino is identical with the anti-neutrino. (In this case,

---

* In the case of a wave function which consists of one component (spin zero), I is simply a number. In the general case the different components of the wave function may be intermixed, and I, in general, is a matrix.

in particular, the magnetic moment of the neutrino must be identically zero.) There is a convincing method for deciding this question. This method is based on measurements of the lifetime for double β-decay. In principle, double β-decay takes place between two isobars which differ in atomic number by two units. The existence of a large number of such isobars in nature (with even numbers of protons and neutrons and spin zero) is due to the fact that either the intermediate nucleus (with odd numbers of protons and neutrons) has a large mass and ordinary β-decay is impossible energetically, or it has high spin so that β-decay, while possible in a formal sense, is actually forbidden to such an extent that it becomes virtually unobservable.

The usual mode of double β-decay can be written schematically in the form

$$2n \rightarrow 2p + 2e^- + \nu + \bar{\nu}.$$

If, however, the neutrino and anti-neutrino were identical, the decay of one neutron could occur with the emission of an anti-neutrino while the decay of the other neutron could occur with the *emission* of a neutrino or, what is essentially the same thing, with the *absorption* of the available anti-neutrino. Thus, the double β-decay process could take place without a neutrino at all: we would have the scheme

$$2n \rightarrow 2p + 2e^-.$$

The theory indicates that the first process is approximately 10,000 times less probable than the second.* Since the second process (without neutrino) has a half-life greater than 10^{16} years for real nuclei, the first mode of decay (anti-neutrino), in general, is unobservable with present-day experimental techniques. Hence, the observation of a double β-decay is possible only if it occurs according to the simple version in which the neutrino does not figure.

Recently, McCarthy has suggested that he has observed double β-decay ($Ca^{48} \rightarrow Ti^{48}$) with a half-life of approximately 10^{17} years. These results would mean that double β-decay occurs according to the scheme in which the neutrino does not appear and that the neutrino is, correspondingly, a true neutral particle. Thus it would follow that the wave function of the neutrino transforms according to $I^2 = -1$ under reflection of coordinates, i. e., its wave function changes sign upon double reflection.

The considerations given above would have a wider meaning. It follows that all particles in nature would transform according to this scheme since we could not have particles which transform according to $I^2 = 1$ and particles which transform according to $I^2 = -1$. The significant feature of these considerations is the fact that the transformation schemes for all spinors would be determined uniquely by the fact that there was even one true neutral particle.

It is obvious, however, that if double β-decay could occur only with the emission of the anti-neutrino it would still be impossible to draw any final conclu-

---

* This is considered in greater detail in Zeldovich, Lukyanov and Smorodinsky, "Properties of the neutrino and double β-decay," Usp. Fiz. Nauk **54**, 361 (1954).

sions, since this fact, in itself, would not tell whether or not there were true neutral particles in nature (even the neutrino).

Having established the transformation scheme for the spinors (in particular $I^2 = -1$) we could now assert that all spinors can have two parities, corresponding to the two possible values of the root of -1. Whence it follows, for example, that a system consisting of two identical particles with half integral spin and without orbital moment is an odd system: the parity of such a system is determined by the products of the parities of the two particles; the reflection transformation will be the product of two identical transformations I, which as we now know is -1. If, however, the particles have different parity, the system is charge-even. In particular, a complex particle with integral spin which decays into two identical particles with half-integral spin and no orbital moment is necessarily an odd particle.

Similar analyses can be carried out for time parity and charge parity (for neutral particles with spin $^1/_2$), but we will not stop here for this purpose.

An interesting application of the theorem of charge parity occurs in the case of positronium—a bound system consisting of an electron and a positron, which is similar to the hydrogen atom.

In accordance with our definition, positronium is a true neutral particle since under charge conjugation the positron becomes an electron while the electron becomes a positron; consequently, positronium is transformed into itself. The charge parity of positronium is determined by its orbital moment and spin. It can be shown that the charge parity of positronium is $(-1)^{L+S}$, where L is the orbital moment and S is the spin. Since the spin of positronium can be 0 (singlet state) or 1 (triplet state), the parity of positronium is the same as the parity determined by L in the singlet state and opposite to the parity determined by L in the triplet state.

We now consider the decay of positronium into photons. We have already indicated that the photon is charge-odd. This means that a system of an odd number of photons is charge-odd. Whence we have the following selection rule: positronium can decay into an even number of photons only if it is in the states ^1S, ^3P, ^1D, etc.; it can decay into an odd number of photons in the states ^3S, ^1P, ^3D, etc.

The positronium states with the lowest energies are ^1S and ^3S. These states are called para-positronium and ortho-positronium, respectively. We have seen that para-positronium (the ^1S state) can decay only into an even number of photons, while ortho-positronium (the ^3S state) can decay only into an odd number. Since decay to one quantum is impossible because of conservation of energy and momentum, we can conclude that the most probable decay schemes for positronium are

$$^1S_0 \rightarrow 2\gamma, \; ^3S_1 \rightarrow 3\gamma.$$

It follows that the decay of ortho-positronium will be less probable than the decay of para-positronium. The experimental work which has been carried out gives the half-lives for these decays as $7 \cdot 10^{-8}$ and $2 \cdot 10^{-10}$ sec, respectively (in agreement with theory).

It is interesting to note the fact that the decay of ortho-positronium into two photons is forbidden as a consequence of only one theorem (namely, the theorem that a system of two photons cannot have a total moment equal to unity).

All the general considerations which we have derived refer to any particle; in particular they refer to all mesons which have been recently discovered.

In these lectures we are considering in detail only the properties of one kind of meson—the π-meson, since its properties have been investigated most completely. This situation is a result of the fact that π-mesons are the only mesons of which intense beams have been obtained in the laboratory with accelerators.*

π-mesons were originally discovered by Powell in cosmic rays and it was only later that they were produced artificially in large accelerators. Three types of π-mesons are produced in collisions of high-energy nucleons with protons or complex nuclei: charged (π^+ and π^-) mesons and neutral (π^0) mesons. The properties of all π-mesons are very similar.

The similarity between π^+- and π^--mesons is obvious from the production processes

$$p \to n + \pi^+, \quad n \to p + \pi^-.$$

It is clear that transformations of this kind could not occur in a free nucleon, since they would violate conservation of energy and momentum. Processes of this type can occur only in the collision of a nucleon with another nucleon or nucleus.

The π^0-meson is produced in a similar way:

$$p \to p + \pi^0, \quad n \to n + \pi^0.$$

All four processes, as we shall see later, are connected by certain relations which appear as a consequence of the isotopic invariance of π-mesons; these relations allow us to consider all three π-mesons in terms of a single particle in three charge states (just as the proton and neutron are considered in terms of one particle).

π-mesons are unstable particles. They decay in the free state. The decay schemes and lifetimes are different for charged π-mesons and neutral π-mesons. Charged π-mesons decay into a μ-meson and a neutrino

$$\pi^{\pm} \to \mu^{\pm} + \nu.$$

Since the mass of the μ-meson is 206 m_e, the decay of the π-meson is characterized by the appearance of an energy of 35 Mev. Because of conservation of momentum the μ-meson acquires 5 Mev while the remaining 30 Mev is carried away by

---

* In recent years μ-mesons have become available in the laboratory as products of the decay of π-mesons. However, the investigation of the properties of these particles is just starting. In 1955, the Bevatron at Berkeley, which yields protons with energies of more than 6 Bev, was able to furnish a beam of K-mesons—particles with a mass equal to 970 electron masses. Only preliminary reports of these experiments have appeared so far.

the neutrino. The half-period of $\pi^{\pm}$-mesons is

$$T_{1/2}(\pi^{\pm}) = (3.6 \pm 0.3)10^{-8} \text{ sec.}$$

The neutral π^0-meson decays into two photons

$$\pi^0 \rightarrow \gamma + \gamma;$$

the half-period for this process is

$$T_{1/2}(\pi^0) = (5 \pm 3)10^{-15} \text{ sec.}$$

It should not be thought that the difference in decay modes and lifetimes violates the charge symmetry between $\pi^{\pm}$-mesons and π^0-mesons. The decay into two photons has no analog in the charged particle case. However, the process $\pi^0 \rightarrow \mu^0 + \nu$, if it exists (up to the present time there has been no indication of the existence of the μ^0-meson), should have a probability approximately 10^6 times smaller than the probability of decay into two photons.

The lifetime of the π^0-meson is itself short compared with the measured lifetimes of elementary particles. The path traversed by the π^0-meson in its lifetime is of the order of 10^{-5} cm. This time, however, is large enough so that we can still speak of the π^0-meson as an independent particle. The characteristic time of a particle (in order of magnitude) is $\hbar/\mu c^2$, since this is the single combination having the dimensions of time which can be formed from the atomic constants. In the case of the π^0-meson this time is $5 \cdot 10^{-24}$ sec; in comparison with this quantity, the lifetime of the π^0-meson is very large. Hence, the reality of the π^0-meson is as well founded as that of any other particle.

With the exception of decay processes the similarities in the properties of the three mesons are very marked.   Leaving out electric effects, which are important only in the study of slow mesons, the production of different π-mesons and the scattering of these particles on nucleons and nuclei are very similar.

This similarity can be noticed immediately from the masses of π-mesons

$$m(\pi^{\pm}) = (272 \pm 0.3) \quad \text{electron masses,}$$

$$m(\pi^0) = (263.9 \pm 0.9) \text{ electron masses.}$$

Although the relative difference in mass is larger than in the case of the neutron and proton (approximately 0.15%), it is small compared with the mass of the meson itself. The similarity in the behavior of π-mesons at the present time has been taken to mean that we can apply the idea of charge invariance to π-mesons and consider them as parts of a triad.

In terms of isotopic space the existence of three states means that the isotopic spin of the π-meson is 1 and that it can have three different projections on the ζ-axis: $+1, 0$ and -1. The three projections of the isotopic spin on the ζ-axis correspond to the three different π-mesons:

$$\tau_{\zeta} = +1 \quad (\pi^+), \qquad \tau_{\zeta} = 0 \quad (\pi^0), \qquad \tau_{\zeta} = -1 \quad (\pi^-).$$

These conditions correspond to those used earlier in choosing the isotopic spin to be associated with the proton as $+1/2$. Naturally, the isotopic invariance hypothesis for π-mesons requires experimental verification.

It is immediately obvious that considerations of charge symmetry can yield certain relations between the cross sections; for example, it may be asserted that in the collision of a proton and a neutron the number of charged mesons (of the same energy and in the same solid angle) of both signs must be equal

$$\sigma(p + n \rightarrow n + n + \pi^+) = \sigma(p + n \rightarrow p + p + \pi^-).$$

Relations of this kind are essentially trivial. However, isotopic invariance leads to a more interesting relation.

At the present time only one relation of this kind has been verified experimentally. There is a well-known reaction in which a deuteron and π^+-meson are produced in the collision of two protons

$$p + p \rightarrow \pi^+ + d.$$

We will encounter this reaction again later on. An analogous reaction is observed in the collision of a neutron with a proton:

$$n + p \rightarrow \pi^0 + d.$$

It can be shown that charge invariance implies that the cross section for the first of these reactions must be twice as large as the cross section for the second:

$$\sigma(p + p \rightarrow \pi^+ + d) = 2\sigma(n + p \rightarrow \pi^0 + d).$$

This relation is easily established if one notes that the final products of the reaction have a total isotopic spin equal to 1 ($T_\pi = 1$, $T_d = 0$). The isotopic spin of the two protons is 1, the isotopic spin of the n-p system is either 0 or 1 (the projection $T_\zeta = 0$). Both of these values are equally probable. Hence, the reaction in question is possible in only half the states of the n-p system. This relation has been verified experimentally.

Interesting relations arise in the production of mesons in the collision of nucleons with nuclei whose isotopic spin is 0. Such nuclei are H^2, He^4, C^{12}, and all nuclei with an even number of protons and neutrons (in the ground state). All three types of mesons can be produced in these collisions. The following relation obtains between the meson-production cross sections:

$$\sigma^+ + \sigma^- = 2\sigma^0;$$

this relation applies for a given meson energy and angle. The relation has not as yet been verified experimentally.

A further problem is that of the spin of π-mesons. Direct measurements of the spin of these particles are extremely difficult. There is, however, a simple and extremely convincing method of indirect measurement. This method is based on an investigation of the relations between the reaction $p + p \rightarrow \pi^+ + d$ and its inverse—the capture of a positive meson by a deuteron.

Using the principle of detailed balance we can establish a relation between these reactions which is independent of any assumption as to the mechanism involved. If we use the symbols p_p and p_π to denote the momentum of the proton and π-meson in the center-of-mass system, and the symbol s to denote the spin of the π-meson

$$\sigma(p + p \rightarrow \pi^+ + d) = {}^3/_2(2s + 1)(\text{p}_p/\text{p}\pi)^2\sigma(d + \pi^+ \rightarrow p + p).$$

Because of the factor $(2s + 1)$, the ratio between these reactions will change markedly depending on whether the spin of the meson is zero $(2s + 1 = 1)$ or unity $(2s + 1 = 3)$. The experimental data definitely indicate that the spin is zero.

It can be independently established that, in any case, the spin of the neutral meson is not unity. This statement follows from the fact, mentioned above, that the π^0-meson decays into two photons; as we have already mentioned, a particle with spin 1 (for example, ortho-positronium) cannot decay into two photons.

The next characteristic of the π-meson which must be established is its parity under coordinate reflection. The parity can be established from the reaction in which a *negative* meson is captured by a deuteron

$$\pi^- + d \rightarrow n + n.$$

This reaction has a high probability and is easily observed. When a π^--meson enters a medium in which there are deuterium nuclei it is slowed down and captured by the deuteron in the K-orbit, forming a neutral system similar to hydrogen. Since capture of the π-meson takes place in the K-orbit, our analysis of the problem is simplified since it is known beforehand that the momentum of this system in the initial state is 1 (deuteron spin). Since the intrinsic parity of the deuteron is the same as the intrinsic parity of a system composed of two neutrons,* the parity of the system is equal to the intrinsic parity of the π-meson.

We consider the final state of the system. A system of two nucleons can have a total moment equal to 1 in the four following states: ${}^3S_1$, ${}^3P_1$, ${}^1P_1$, ${}^3D_1$. Of these states, only one is available to two identical nucleons, i. e., the ${}^3P_1$ state. Thus, the parity of the system is determined uniquely. The state is odd; consequently, the π^--meson is an odd particle.

It is reasonable to assume that the π^0-meson is also an odd particle. In principle, there is a possibility for a direct measurement of the absolute parity of the π^0-meson, in contrast with the parity determination of the $\pi^\pm$-mesons. The theory indicates that if the π^0-meson is even it will decay into two photons, with parallel polarization (in the system in which the π^0-meson is at rest); if the π^0-meson is odd, the polarization planes for the two photons are mutually perpendicular. This experiment has not been carried out because of the serious experimental difficulties involved.

Usually, in speaking of the parity of mesons with respect to coordinate reflection, instead of the designation "odd π-meson" it is customary to say that the meson is a pseudoscalar particle, since an odd particle with spin zero is described by a one-component wave function which changes sign under coordinate reflection and is therefore a pseudoscalar.

It is also easy to establish the fact that π-mesons are even particles under time

---

* We consider the intrinsic parity of the proton and neutron to be the same, thus determining the parity of the π-meson with respect to the *p-n* system.

reflection.   This result is obvious directly from an examination of the processes in which these particles are formed in collisions between two nucleons.   In such processes the number of nucleons remains unchanged; consequently, their time parity remains unchanged.   Hence, the π-meson is even with respect to time reflection.

Finally, from the fact that the π^0-meson decays into two photons we can conclude that the π^0-meson is an even particle under charge conjugation; that is to say, its wave function is not changed under charge conjugation.

Interaction of π-Mesons with Nucleons

The problem which arises, once the basic properties of π-mesons have been established, is the examination of the laws which govern the interaction of these particles with nucleons. It turns out that the situation here is fundamentally different from that which obtains in electrodynamics. In particular, in contrast with the interaction of an electron with the electromagnetic field, the interaction of mesons with nucleons is "strong" rather than "weak."

The strength of the interaction between the electron and the electromagnetic field is determined by the magnitude of the charge. We must not forget in discussing the numerical value of any physical quantity in theoretical physics that the value depends on which particular set of units is used to express the measurement. Hence, the notion of a "large quantity" or "small quantity" can be formulated only when we indicate with what the quantity in question is being compared; when this concept is applied to dimensionless quantities which are independent of a choice of units, a comparison is made with unity.

Electric charge is a quantity with dimensions. However, we can form a dimensionless constant from the electric charge and other constants which appear in the quantum mechanical equations:

$$\alpha = \frac{e^2}{\hbar c}.$$

No other dimensionless constants can be formed from the parameters of a charged particle and the electromagnetic field; the constant α then can be taken as a measure of the strength of the interaction of the charged particle and the electromagnetic field. As is well known, its magnitude is small and is numerically approximately $1/137$. In many cases a still smaller quantity is important: $e^2/2\pi\hbar c \sim 1/1000$. Hence electrodynamics is an example of a weak-coupling theory.

Because the quantity $e^2/\hbar c$ is so small, it is possible, using present-day perturbation-theory techniques, to calculate any electrodynamic effect to any approximation. In recent years there have been brilliant demonstrations of the agreement between experiment and the theoretical calculations of the corrections to the energy levels in the hydrogen atom (the so-called Lamb shift) and the calculations of the intrinsic magnetic moment of the electron, which is $e\hbar/2mc$ in the first approximation; in the second approximation it is

$$\frac{e\hbar}{2mc}\left(1 + \frac{\alpha}{2\pi}\right).$$

In contrast with the weak coupling which is characteristic of electrodynamics, it turns out that meson interactions are strong.

The clearest indication of the strong interaction appears in the production of π-mesons by photons in nuclei (*meson photoeffect*). If heavy nuclei are bombarded by photons with sufficiently high energy, the cross section for the production of π-mesons does not increase in proportion to the number of particles in the nucleus, i. e., in proportion to the atomic weight A; it goes approximately as $A^{2/3}$.

It is apparent that the probability for the production of mesons should be proportional to the number of particles in the nucleus—the interaction between the photon and nucleus is small and the probability that a photon is absorbed in the nucleus is small; hence, the photon intensity is uniform over the nucleus and the nucleons absorb independently. However, not all the mesons which are produced in the nucleus can escape. Because of the strong meson interaction there is a high probability that mesons are absorbed by one of the nucleons in the nucleus; only a fraction of the mesons escape: namely, those formed near the surface of the nucleus. It is obvious that in this case the number of mesons should be proportional to the nuclear surface area, i. e., proportional to $A^{2/3}$. Thus, the observed dependence of the meson photoeffect on atomic weight is an indication of the small meson range in nuclear matter; in turn, this fact is evidence of the existence of a strong interaction.

Perturbation theory cannot be applied in the case of a strong interaction. On the other hand, the practical approach to calculating these effects is based on perturbation theory. In addition, Pomeranchuk has shown that all forms of the interaction which have been considered up to now are useless when examined in detail.

Thus, at the present time we have no theory for meson interactions; moreover it is clear that the formulation of such a theory will be impossible without radical modification of some of the fundamental concepts.

Notwithstanding the absence of any quantitative meson theory there is no doubt that π-mesons play an important role in nuclear interactions and are responsible for many of the properties of nucleons.

In our first lecture we have noted that the magnetic moment of the proton is 2.79 (in nuclear magnetons), while the magnetic moment of the neutron is -1.91. According to the Dirac equation, however, the magnetic moments of these nucleons should be 1 and 0, respectively.

Hence, the proton has an excess positive magnetic moment of 1.79, while the neutron has an excess negative moment of 1.91. The fact that these two numbers are so close in value is far from accidental. There is little doubt that this additional magnetic moment results from the interaction of the nucleon with the meson field. Charge invariance considerations allow us to state that the magnetic moments associated with the mesons should be the same in magnitude and opposite in sign in these nucleons since the proton and π^--meson form a pair similar to the neutron and π^+-meson. However, the lack of any theory makes it impossible to give a quantitative analysis of this effect.

It is also undoubtedly true that π-mesons play an important role in nuclear forces. From the point of view of contemporary theoretical physics the interaction of protons and neutrons can be described in terms of meson exchange. If, for example, we have a proton and neutron, the proton can emit a π^+-meson, thereby becoming a neutron, while the emitted meson can be absorbed by the neutron, which is then transformed into a proton. This meson exchange automatically leads to nucleon interaction, since radiation and absorption (even virtual radiation and absorption) are accompanied by a change in the nucleon momentum. In the exchange of charged mesons the charge of the particle is changed; as a result the proton becomes a neutron and the neutron becomes a proton, and the interaction is of exchange character. On the other hand, the exchange of neutral mesons leads to the usual (nonexchange) forces; thus the meson mechanism provides for both types of forces.

To estimate the range of the forces which are to be associated with meson exchange we must return to dimensionality considerations. The range can depend only on the meson mass. The mass of the nucleon does not appear in an important way in the range expression since this range must be finite for a nucleon at rest (infinitely heavy nucleon). Using the mass of the π-meson (μ) and the constants $\hbar$ and c we can form only one constant which has the dimensions of length—the Compton wavelength of the meson*

$$r_0 \sim \frac{\hbar}{\mu c} \simeq 1.4 \cdot 10^{-13} \text{ cm.}$$

This quantity then determines the range of the interaction. The dimensional estimate, as we have seen, agrees with the range of nuclear forces ($\sim 10^{-13}$ cm), thereby tending to support the mesonic origin of these forces.

It is possible that the heavier mesons which have been discovered in recent years may also make a contribution to nuclear forces. However, the experimental data on these particles are as yet very incomplete, so that it is impossible to say anything more definite about them.

We now consider the existing experimental data on the interaction of π-mesons with nucleons. Of the enormous amount of experimental data which is available we shall be concerned only with those effects which allow us to draw direct conclusions as to properties of π-mesons.

We start with the simplest process: the scattering of π-mesons by nucleons. This process has been studied over a wide meson-energy range: from 20 to 1500 Mev; research has been done on the scattering of both π^+ and π^--mesons by protons. Meson charge exchange ($\pi^- + p \to \pi^0 + n$) has also been studied up to several hundreds of millions of electron volts.

The energy dependence of the cross section for these reactions is shown in Fig. 9. The cross section for the scattering of π^+ on protons (equal to the cross

---

* If the mass of the meson were zero (as is that of the photon) it would be impossible to form a constant with the dimensions of length. In this case the forces would have an infinite range (like Coulomb forces).

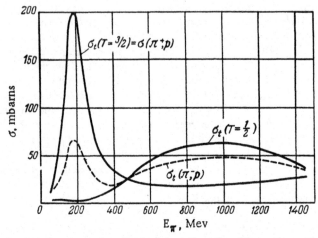

Fig. 9. Total cross sections for the interaction of positive and negative π-mesons with protons (the figure is from an unpublished survey by Barkov and Nikolsky). The interaction curves in the states with isotopic spin $T = 1/2$ are obtained from the experimental data using the formula $2\sigma_t(T = 1/2) = 3\sigma_t(\pi^-) - \sigma_t(\pi^+)$; E_π is the energy of the π-mesons in the laboratory system.

section for π^- on neutrons) reaches a maximum of approximately 200 mbarns at an energy of 200 Mev. The total cross section for π^- scattering (scattering and charge exchange) on protons also reaches a maximum in this region, but its magnitude is smaller, approximately 65 mbarns; approximately $1/3$ of this cross section is due to elastic scattering. As the energy is increased the cross section falls off rapidly (at an energy of 360 Mev, $\sigma_t(\pi^+, p) = 50$ mbarns) while at the highest energies which have been achieved, 1500 Mev, the cross sections are approximately 35 mbarns for $\sigma_t(\pi^-, p)$ and 30 mbarns for $\sigma_t(\pi^+, p)$.

It is interesting to note that if only the state with isotopic spin $3/2$ participates in scattering* the scattering cross sections should obey the following relation:

$$\sigma(p, \pi^+ \rightarrow p, \pi^+) : \sigma(p, \pi^- \rightarrow p, \pi^0) : \sigma(p, \pi^- \rightarrow p, \pi^-) = 9 : 2 : 1.$$

We see that this relation is approximately satisfied in the region of the maximum (at higher energies the cross sections are almost the same). Thus, in this region the interaction is a strong function of the isotopic spin—a property which we have already noted in neutron interactions.

The pattern seems to be associated with scattering in the "level" with $T = 3/2$, and the phase analysis indicates that the corresponding state is $p_{1/2}$.

It is extremely interesting to examine the cross section for the meson photoeffect in the proton. This effect is also characterized by a maximum cross section at a photon energy of approximately 200 Mev, indicating a strong inter-

---

* Obviously, the π-meson and the nucleon can have a total isotopic spin of either $1/2$ or $3/2$.

action in the $T = {}^3/_2$ state; unfortunately the absence of any theory prevents our carrying out a more thorough analysis of the experimental data.

Interesting effects are observed when mesons are produced as a result of nucleon collisions. In proton collisions, both π^+- and π^0-mesons can be produced in principle:

$$p + p \rightarrow p + n + \pi^+,$$

$$p + p \rightarrow p + p + \pi^0.$$

Actually, however, the cross section for the production of π^0-mesons is very small and there are virtually no π^0-mesons near threshold. Thus, at an energy of 460 Mev the cross section is divided between three possible reactions, as follows:

elastic (p, p)	20	mbarns
π^+ production	4.5	mbarns
π^0 production	0.4	mbarns
Total cross section ~ 25		mbarns

Even at 660 Mev the fraction of π^0-mesons produced in p-p collisions amounts to only 10% of the total cross section.

The small probability for the production of π^0-mesons can be explained if we assume that mesons are produced in singlet states of the p-p system. Since there is little excess energy close to threshold, the final state will most probably have $L = 0$ (${}^3S_1$ or ${}^1S_0$); the two protons can occupy only the ${}^1S_0$ state, but both states are available to a proton and neutron. It is easy to see that the reaction is forbidden for two protons in a final state ${}^1S_0$ if the initial states are singlet states. The singlet states of the p-p system (${}^1S_0$, ${}^1D_2$) have even momenta and are even. Since the final state of the p-p system has zero moment and is even, all the initial momentum is transferred to the meson. But the meson is odd; hence, having even moment it must be in an odd state. Consequently, the process is forbidden by conservation laws.

There is one further restriction on the production of mesons in neutron-proton collisions. The following reactions are possible:

$$n + p \rightarrow p + p + \pi^-,$$

$$n + p \rightarrow n + p + \pi^0.$$

The first reaction has a rather small probability. This effect can be explained in the same way if it is assumed that mesons are produced only in singlet states of the colliding nucleons.

As the energy increases the meson-production cross section also increases. At 850 Mev the π-meson production cross section in proton-proton interactions reaches a value of approximately 25 mbarns and remains more or less constant up to 1275 Mev. As we have already indicated, this cross section coincides with the elastic cross section and corresponds to the interaction of two absolutely absorbing spheres—protons with a radius of $4.5 \cdot 10^{-14}$ cm.

The extent of the energy region for which this simple model applies is as yet unknown.   It is interesting that π-meson scattering on protons at 1400 Mev can also be described by a model in which the proton is considered a "black" sphere with the same radius.*

There is one more effect, first observed by Migdal,† which is observed in meson production in nucleon-nucleon collisions near threshold.   The meson spectrum indicates that the mesons carry away a large part of the energy so that the fraction left to the nucleons is relatively small.   This energy distribution is related to the nature of the nucleon interaction.   We have seen that this inter-action exhibits a resonance effect at low energies; it follows, as can be demon-strated rigorously, that the nucleons have high probability of having small relative velocities (strong interaction).

This effect is so strong that deuteron formation is very probable in the final state; at these high energies this effect is very strange.   At an energy of 460 Mev a deuteron is formed in approximately $1/_3$ of the cases in which π^+-mesons are produced (in proton-proton collisions); at 660 Mev the number of deuterons is still sizable.   Deuteron formation due to resonance effects is observed not only in meson production but also in other nuclear reactions.

As the nucleon energy available to experimenters has increased, new effects have been observed.   At the present time experiments are reported in which studies have been made of events in which several π-mesons are produced in meson-meson interactions.   However, as yet these are very rare and an analysis would not be fruitful.

New effects are observed at still higher energies; as the energy is increased the probability of multiple-meson production becomes greater than the probability of single-meson production.

When nucleons with energies greater than 10,000 Mev collide with nuclei, nuclear disintegration events ("stars") occur in which the majority of particles are mesons.   The cross section for star production is comparable with the geometric dimension of the nucleus—πR^2 (R is the nuclear radius).

This property of the interaction distinguishes the "strong" nuclear interaction from the "weak" electromagnetic interaction.   The formation of two photons of comparable energy is always less probable (by a factor of α) than the formation of a single photon;‡ the cross section for the formation of three photons is still less probable.   No matter how high the energy, there is never multiple produc-tion of high-energy photons solely as a result of an electromagnetic interaction.§ On the other hand, the intensity of the nuclear interaction increases with energy, as does the number of π-mesons produced in collisions.

---

* W. B. Fowler, R. M. Lee, W. D. Shephard, R. P. Shutt, A. M. Thorndike and A. Winston, Phys. Rev. **97**, 809 (1955).

† A. B. Migdal, J. Exptl.-Theoret. Phys. **28**, 3 (1955).

‡ It should be emphasized, however, that this statement does not apply to low-energy photons.   As is indicated by the theory, any process is accompanied by the radiation of photons with very low energy; these do not play any significant role in the majority of cases.   This effect is called the infrared catastrophe.

§ In stars many photons arise as a result of π-meson decay; the production of these mesons is of multiple character.

In the absence of any meson theory we cannot offer any analysis of multiple-meson production; the extent to which the single-meson production mechanism applies in this case is also not clear.

The fact that the interaction increases with energy, however, does allow us to formulate a quantitative theory for the effects in the high energy region, in which the number of mesons becomes so large that thermodynamic considerations can be applied. The idea of applying thermodynamics in multiple production of mesons was first suggested by Fermi.* The method proposed by Fermi is based on the assumption that in the collision of high-energy nucleons all the energy appears instantaneously in a small volume and that both nucleons come to rest in the center-of-mass system.

Because of the Lorentz contraction this volume is not spherical but is compressed in the direction of motion of the nucleons prior to the collision. The amount of compression is given by the well-known factor

$$\sqrt{1 - \frac{v^2}{c^2}} = \frac{Mc^2}{E'},$$

where M is the mass of the nucleon and E' its energy in the center-of-mass system. The latter is related to the energy of the incoming nucleon E by the relation

$$E = \frac{2E'^2}{Mc^2}.$$

The volume of the region in which energy appears is (order of magnitude)

$$\Omega \cong \frac{Mc^2}{E'} \left(\frac{\hbar}{\mu c} \right)^3,$$

since $\hbar/\mu c$ (μ is the meson mass) is of the order of the interaction range. We assume that the interaction is so strong that there is time for statistical equilibrium to be established during the collision; thus, the entire system can be characterized by a temperature T_0 (which, of course, is very high).

If T_0 is so large that

$$T_0 \gg \mu c^2,$$

many mesons can be produced in the system. However, the conditions under which we may speak of a definite number of mesons in a system are not clear; at high densities, when the dimensions of the system and the interaction range are of the same order of magnitude the notion of what we mean by individual particles, as far as the principles of quantum mechanics are concerned, is not obvious. Hence, one does not expect the number of mesons observed in a meson star to be determined by the equilibrium conditions in the initial stages of the collision.

---

* A translation of the Fermi paper is given in Usp. Fiz. Nauk **46**, 71 (1952). A detailed presentation of these problems is given in a paper by S. Z. Belenky and L. D. Landau, Usp. Fiz. Nauk **56**, 309 (1955).

During the entire time in which the system is in a highly compressed state, new particles are being produced and old ones are being absorbed. These processes come to a halt only when the system has expanded to such an extent that the interactions become small. The number of particles which remain in the system at this time will be different from the number of particles at the initial stage of the collision and will be more or less determined by the expansion process. Hence, it is obvious that thermodynamic considerations alone are not sufficient for a description of the system; we must also be concerned with the hydrodynamics involved in the expansion of a system in thermal equilibrium. This problem, however, is not analogous to the classical hydrodynamic problem since the velocity of the particles is very high and relativistic equations must be used.

There are certain general considerations, derived from relativity theory, which allow us to draw conclusions as to the dependence of the number of particles on the energy of the initial incoming particle. In this problem an important part is played by the equation of state of the system—the relation between the energy density and the pressure. This equation cannot be introduced in general form—it is determined by the actual properties of the system and is different for different systems. However, using relativistic mechanics, we can indicate a certain limiting form for the equation of state. Specifically, it can be shown that the energy density ε and the pressure p must satisfy the inequality

$$p \leqslant \varepsilon/3.$$

As is known, in the case of light the relation $p = \varepsilon/3$ applies; the same relation is valid for an ideal relativistic gas (i. e., a gas of noninteracting particles moving with high velocity). It may be assumed that in the case of a highly interacting gas the pressure is close to the limiting value; hence, we shall postulate that the equation of state for the system being considered is of the form

$$p = \varepsilon/3.$$

Now, in considering the rapid dispersion of the particles we must take account of the fact that the temperature of the system is reduced as the particles fly apart. As the temperature decreases the density of the particles decreases, and there is a resultant reduction in interaction, i. e., the probability of creation and annihilation of particles is also decreased. It is clear that the final number of outgoing particles is determined when the temperature of the system falls to a value such that a further change in the number of particles becomes very improbable. In order-of-magnitude terms this temperature T_f is determined by the relation

$$T_f \sim \mu c^2.$$

The temperature T_f should be essentially independent of the properties of the system because the equilibrium density of the particles is a rapidly varying function of temperature. During the period of hydrodynamic expansion, because of the strong interaction between particles the path lengths remain small so that we can neglect viscosity and thermal-conduction effects and consider the hydrodynamic stage of the expansion as an adiabatic process. At the same time, the ratio of the density of particles to the entropy density of the system is a very slowly varying function of temperature. Since the temperature is approximately constant during expansion we may consider this ratio as constant and independent of the properties of the system. Consequently, the entropy density and the density of the number of particles are proportional:

$$n = \text{const } s,$$

where s is in dimensionless units; the constant, from dimensionality considerations, must be of the order of $(\mu c/\hbar)^3$. These relations also allow us to determine the approximate dependence of the number of particles on the energy of the incoming nucleon.

Since we have assumed that the pressure is $1/3$ of the energy density, it follows from well-known thermodynamic relations that

$$s \sim \mathcal{E}^{3/4}.$$

This can be shown most simply by an analogy with black-body radiation, in which the same law applies ($\mathcal{E} = 3p$). In the case of a blackbody $\mathcal{E} \sim T^4$, and consequently

$$T \sim \mathcal{E}^{1/4};$$

on the other hand, for a unit volume

$$d\mathcal{E} = T ds,$$

whence it follows that $\mathcal{E}^{1/4} \sim s$.

As we have indicated, the volume of the system Ω is equal to the volume of a sphere of radius $a \sim (\hbar/\mu c)$, reduced by the factor Mc^2/E'. All the energy which appears in the collision, i. e., $2E'$, appears in the volume of this ellipsoid. Consequently, the energy density is

$$\mathcal{E} = \frac{2E'}{\Omega} = \frac{2E'}{\frac{4}{3}\pi a^3 (Mc^2/E')} \sim E'^2.$$

Therefore, the entropy density is

$$s \sim \mathcal{E}^{3/4} \sim E'^{3/2}$$

(E' is the energy in the center-of-mass system). Since the density of particles is proportional to the entropy density,

$$n \sim E'^{3/2},$$

and the total number of particles is

$$N = n\Omega \sim E'^{3/2} \frac{1}{E'} = E'^{1/2},$$

since Ω is inversely proportional to the energy.

In order to transform to energy in the laboratory system, E, we must substitute $(E')^2 = 1/2 Mc^2 E$. Finally we obtain

$$N \sim E^{1/4}.$$

This expression coincides with the relation obtained by Fermi (although he started from the incorrect assumption that N is the number of particles created at the instant of collision).

A noteworthy feature of this formula is the fact that it predicts a slow growth in creation multiplicity with energy of the incoming particle. This result explains the well-known experimental fact that stars with more than a hundred particles are observed very rarely and that the majority of observed stars have a comparatively small number of high-energy particles (the mean particle energy $E_{mean} \sim E/N \sim E^{3/4}$).

Strictly speaking, collisions which are observed in cosmic rays are not collisions between two nucleons, but collisions between nucleons and nuclei or between nuclei and nuclei. The picture for a collision between a nucleon and a nucleus is virtually the same as that for a collision between two nucleons, except that the nucleon has a radius much smaller than the nucleus, so that it strikes only those nucleons in the nucleus which lie in its path. Hence, the number of particles created in such collision will be essentially the same as the number created in collisions between two nucleons, and will be relatively independent of the atomic weight of the nucleus.

If, however, both nuclei are heavy, the number of created particles will be a strong function of the dimensions of the nucleus. In this case the number of created particles is proportional to the atomic weight raised to the three-fourths power:

$$N \sim E^{1/4} A^{3/4}.$$

Thus, a nucleus turns out to be much more effective, in terms of creation of new particles, than a proton with the same energy.

We may also consider the angular distribution of the created particles. Fermi assumed that the particles are emitted isotropically in the center-of-mass system. However, there is no basis for this assumption. A careful analysis of the angular distribution requires a complete solution of the rather complicated hydrodynamic problem; qualitative considerations alone are not very useful.

This solution can, in fact, be obtained. It leads to the following picture of the angular distribution. In the center-of-mass system the majority of particles are emitted at large angles with respect to the direction of the incoming particle. However, the density of particles at small angles is larger, so that the angular distribution curve is highly anisotropic. In the laboratory system the distribution also has a peak at angles close to $0°$.

This type of angular distribution leads to a peculiar energy distribution. In this distribution a comparatively large number of particles have energies higher than the mean energy and a large number of particles have energies considerably lower than the mean value.

A comprehensive comparison of the statistical theory of multiple production of π-mesons with experiments cannot be carried out since only a few stars have been recorded, and these are not suitable for analysis.

A CATALOG OF SELECTED
DOVER BOOKS
IN SCIENCE AND MATHEMATICS

Astronomy

BURNHAM'S CELESTIAL HANDBOOK, Robert Burnham, Jr. Thorough guide to the stars beyond our solar system. Exhaustive treatment. Alphabetical by constellation: Andromeda to Cetus in Vol. 1; Chamaeleon to Orion in Vol. 2; and Pavo to Vulpecula in Vol. 3. Hundreds of illustrations. Index in Vol. 3. 2,000pp. 6⅛ x 9¼.

Vol. I: 0-486-23567-X
Vol. II: 0-486-23568-8
Vol. III: 0-486-23673-0

EXPLORING THE MOON THROUGH BINOCULARS AND SMALL TELE-SCOPES, Ernest H. Cherrington, Jr. Informative, profusely illustrated guide to locating and identifying craters, rills, seas, mountains, other lunar features. Newly revised and updated with special section of new photos. Over 100 photos and diagrams. 240pp. 8¼ x 11. 0-486-24491-1

THE EXTRATERRESTRIAL LIFE DEBATE, 1750–1900, Michael J. Crowe. First detailed, scholarly study in English of the many ideas that developed from 1750 to 1900 regarding the existence of intelligent extraterrestrial life. Examines ideas of Kant, Herschel, Voltaire, Percival Lowell, many other scientists and thinkers. 16 illustrations. 704pp. 5⅜ x 8½. 0-486-40675-X

THEORIES OF THE WORLD FROM ANTIQUITY TO THE COPERNICAN REVOLUTION, Michael J. Crowe. Newly revised edition of an accessible, enlightening book re-creates the change from an earth-centered to a sun-centered conception of the solar system. 242pp. 5⅜ x 8½. 0-486-41444-2

ARISTARCHUS OF SAMOS: The Ancient Copernicus, Sir Thomas Heath. Heath's history of astronomy ranges from Homer and Hesiod to Aristarchus and includes quotes from numerous thinkers, compilers, and scholasticists from Thales and Anaximander through Pythagoras, Plato, Aristotle, and Heraclides. 34 figures. 448pp. 5⅜ x 8½. 0-486-43886-4

A COMPLETE MANUAL OF AMATEUR ASTRONOMY: TOOLS AND TECHNIQUES FOR ASTRONOMICAL OBSERVATIONS, P. Clay Sherrod with Thomas L. Koed. Concise, highly readable book discusses: selecting, setting up and maintaining a telescope; amateur studies of the sun; lunar topography and occultations; observations of Mars, Jupiter, Saturn, the minor planets and the stars; an introduction to photoelectric photometry; more. 1981 ed. 124 figures. 25 halftones. 37 tables. 335pp. 6½ x 9¼. 0-486-42820-8

AMATEUR ASTRONOMER'S HANDBOOK, J. B. Sidgwick. Timeless, comprehensive coverage of telescopes, mirrors, lenses, mountings, telescope drives, micrometers, spectroscopes, more. 189 illustrations. 576pp. 5⅜ x 8¼. (Available in U.S. only.) 0-486-24034-7

STAR LORE: Myths, Legends, and Facts, William Tyler Olcott. Captivating retellings of the origins and histories of ancient star groups include Pegasus, Ursa Major, Pleiades, signs of the zodiac, and other constellations. "Classic."—Sky & Telescope. 58 illustrations. 544pp. 5⅜ x 8½. 0-486-43581-4

Chemistry

THE SCEPTICAL CHYMIST: THE CLASSIC 1661 TEXT, Robert Boyle. Boyle defines the term "element," asserting that all natural phenomena can be explained by the motion and organization of primary particles. 1911 ed. viii+232pp. $5^3/_8$ x $8^1/_2$.
0-486-42825-7

RADIOACTIVE SUBSTANCES, Marie Curie. Here is the celebrated scientist's doctoral thesis, the prelude to her receipt of the 1903 Nobel Prize. Curie discusses establishing atomic character of radioactivity found in compounds of uranium and thorium; extraction from pitchblende of polonium and radium; isolation of pure radium chloride; determination of atomic weight of radium; plus electric, photographic, luminous, heat, color effects of radioactivity. ii+94pp. $5^1/_8$ x $8^1/_2$.
0-486-42550-9

CHEMICAL MAGIC, Leonard A. Ford. Second Edition, Revised by E. Winston Grundmeier. Over 100 unusual stunts demonstrating cold fire, dust explosions, much more. Text explains scientific principles and stresses safety precautions. 128pp. $5^3/_8$ x $8^1/_2$.
0-486-67628-5

MOLECULAR THEORY OF CAPILLARITY, J. S. Rowlinson and B. Widom. History of surface phenomena offers critical and detailed examination and assessment of modern theories, focusing on statistical mechanics and application of results in mean-field approximation to model systems. 1989 edition. 352pp. $5^3/_8$ x $8^1/_2$.
0-486-42544-4

CHEMICAL AND CATALYTIC REACTION ENGINEERING, James J. Carberry. Designed to offer background for managing chemical reactions, this text examines behavior of chemical reactions and reactors; fluid-fluid and fluid-solid reaction systems; heterogeneous catalysis and catalytic kinetics; more. 1976 edition. 672pp. $6^1/_8$ x $9^1/_4$.
0-486-41736-0 $31.95

ELEMENTS OF CHEMISTRY, Antoine Lavoisier. Monumental classic by founder of modern chemistry in remarkable reprint of rare 1790 Kerr translation. A must for every student of chemistry or the history of science. 539pp. $5^3/_8$ x $8^1/_2$.
0-486-64624-6

MOLECULES AND RADIATION: An Introduction to Modern Molecular Spectroscopy. Second Edition, Jeffrey I. Steinfeld. This unified treatment introduces upper-level undergraduates and graduate students to the concepts and the methods of molecular spectroscopy and applications to quantum electronics, lasers, and related optical phenomena. 1985 edition. 512pp. $5^3/_8$ x $8^1/_2$.
0-486-44152-0

A SHORT HISTORY OF CHEMISTRY, J. R. Partington. Classic exposition explores origins of chemistry, alchemy, early medical chemistry, nature of atmosphere, theory of valency, laws and structure of atomic theory, much more. 428pp. $5^3/_8$ x $8^1/_2$. (Available in U.S. only.)
0-486-65977-1

GENERAL CHEMISTRY, Linus Pauling. Revised 3rd edition of classic first-year text by Nobel laureate. Atomic and molecular structure, quantum mechanics, statistical mechanics, thermodynamics correlated with descriptive chemistry. Problems. 992pp. $5^3/_8$ x $8^1/_2$.
0-486-65622-5

ELECTRON CORRELATION IN MOLECULES, S. Wilson. This text addresses one of theoretical chemistry's central problems. Topics include molecular electronic structure, independent electron models, electron correlation, the linked diagram theorem, and related topics. 1984 edition. 304pp. $5^3/_8$ x $8^1/_2$.
0-486-45879-2

Engineering

DE RE METALLICA, Georgius Agricola. The famous Hoover translation of greatest treatise on technological chemistry, engineering, geology, mining of early modern times (1556). All 289 original woodcuts. 638pp. 6¾ x 11. 0-486-60006-8

FUNDAMENTALS OF ASTRODYNAMICS, Roger Bate et al. Modern approach developed by U.S. Air Force Academy. Designed as a first course. Problems, exercises. Numerous illustrations. 455pp. 5⅜ x 8½. 0-486-60061-0

DYNAMICS OF FLUIDS IN POROUS MEDIA, Jacob Bear. For advanced students of ground water hydrology, soil mechanics and physics, drainage and irrigation engineering and more. 335 illustrations. Exercises, with answers. 784pp. 6⅛ x 9¼. 0-486-65675-6

THEORY OF VISCOELASTICITY (SECOND EDITION), Richard M. Christensen. Complete consistent description of the linear theory of the viscoelastic behavior of materials. Problem-solving techniques discussed. 1982 edition. 29 figures. xiv+364pp. 6⅛ x 9¼. 0-486-42880-X

MECHANICS, J. P. Den Hartog. A classic introductory text or refresher. Hundreds of applications and design problems illuminate fundamentals of trusses, loaded beams and cables, etc. 334 answered problems. 462pp. 5⅜ x 8½. 0-486-60754-2

MECHANICAL VIBRATIONS, J. P. Den Hartog. Classic textbook offers lucid explanations and illustrative models, applying theories of vibrations to a variety of practical industrial engineering problems. Numerous figures. 233 problems, solutions. Appendix. Index. Preface. 436pp. 5⅜ x 8½. 0-486-64785-4

STRENGTH OF MATERIALS, J. P. Den Hartog. Full, clear treatment of basic material (tension, torsion, bending, etc.) plus advanced material on engineering methods, applications. 350 answered problems. 323pp. 5⅜ x 8½. 0-486-60755-0

A HISTORY OF MECHANICS, René Dugas. Monumental study of mechanical principles from antiquity to quantum mechanics. Contributions of ancient Greeks, Galileo, Leonardo, Kepler, Lagrange, many others. 671pp. 5⅜ x 8½. 0-486-65632-2

STABILITY THEORY AND ITS APPLICATIONS TO STRUCTURAL MECHANICS, Clive L. Dym. Self-contained text focuses on Koiter postbuckling analyses, with mathematical notions of stability of motion. Basing minimum energy principles for static stability upon dynamic concepts of stability of motion, it develops asymptotic buckling and postbuckling analyses from potential energy considerations, with applications to columns, plates, and arches. 1974 ed. 208pp. 5⅜ x 8½. 0-486-42541-X

BASIC ELECTRICITY, U.S. Bureau of Naval Personnel. Originally a training course; best nontechnical coverage. Topics include batteries, circuits, conductors, AC and DC, inductance and capacitance, generators, motors, transformers, amplifiers, etc. Many questions with answers. 349 illustrations. 1969 edition. 448pp. 6½ x 9¼. 0-486-20973-3

ROCKETS, Robert Goddard. Two of the most significant publications in the history of rocketry and jet propulsion: "A Method of Reaching Extreme Altitudes" (1919) and "Liquid Propellant Rocket Development" (1936). 128pp. 5⅜ x 8½.   0-486-42537-1

STATISTICAL MECHANICS: PRINCIPLES AND APPLICATIONS, Terrell L. Hill. Standard text covers fundamentals of statistical mechanics, applications to fluctuation theory, imperfect gases, distribution functions, more. 448pp. 5⅜ x 8½.   0-486-65390-0

ENGINEERING AND TECHNOLOGY 1650–1750: ILLUSTRATIONS AND TEXTS FROM ORIGINAL SOURCES, Martin Jensen. Highly readable text with more than 200 contemporary drawings and detailed engravings of engineering projects dealing with surveying, leveling, materials, hand tools, lifting equipment, transport and erection, piling, bailing, water supply, hydraulic engineering, and more. Among the specific projects outlined-transporting a 50-ton stone to the Louvre, erecting an obelisk, building timber locks, and dredging canals. 207pp. 8⅜ x 11¼.   0-486-42232-1

THE VARIATIONAL PRINCIPLES OF MECHANICS, Cornelius Lanczos. Graduate level coverage of calculus of variations, equations of motion, relativistic mechanics, more. First inexpensive paperbound edition of classic treatise. Index. Bibliography. 418pp. 5⅜ x 8½.   0-486-65067-7

PROTECTION OF ELECTRONIC CIRCUITS FROM OVERVOLTAGES, Ronald B. Standler. Five-part treatment presents practical rules and strategies for circuits designed to protect electronic systems from damage by transient overvoltages. 1989 ed. xxiv+434pp. 6⅛ x 9¼.   0-486-42552-5

ROTARY WING AERODYNAMICS, W. Z. Stepniewski. Clear, concise text covers aerodynamic phenomena of the rotor and offers guidelines for helicopter performance evaluation. Originally prepared for NASA. 537 figures. 640pp. 6⅛ x 9¼. 0-486-64647-5

INTRODUCTION TO SPACE DYNAMICS, William Tyrrell Thomson. Comprehensive, classic introduction to space-flight engineering for advanced undergraduate and graduate students. Includes vector algebra, kinematics, transformation of coordinates. Bibliography. Index. 352pp. 5⅜ x 8½.   0-486-65113-4

HISTORY OF STRENGTH OF MATERIALS, Stephen P. Timoshenko. Excellent historical survey of the strength of materials with many references to the theories of elasticity and structure. 245 figures. 452pp. 5⅜ x 8½.   0-486-61187-6

ANALYTICAL FRACTURE MECHANICS, David J. Unger. Self-contained text supplements standard fracture mechanics texts by focusing on analytical methods for determining crack-tip stress and strain fields. 336pp. 6⅛ x 9¼.   0-486-41737-9

STATISTICAL MECHANICS OF ELASTICITY, J. H. Weiner. Advanced, self-contained treatment illustrates general principles and elastic behavior of solids. Part 1, based on classical mechanics, studies thermoelastic behavior of crystalline and polymeric solids. Part 2, based on quantum mechanics, focuses on interatomic force laws, behavior of solids, and thermally activated processes. For students of physics and chemistry and for polymer physicists. 1983 ed. 96 figures. 496pp. 5⅜ x 8½.   0-486-42260-7

Mathematics

FUNCTIONAL ANALYSIS (Second Corrected Edition), George Bachman and Lawrence Narici. Excellent treatment of subject geared toward students with background in linear algebra, advanced calculus, physics and engineering. Text covers introduction to inner-product spaces, normed, metric spaces, and topological spaces; complete orthonormal sets, the Hahn-Banach Theorem and its consequences, and many other related subjects. 1966 ed. 544pp. 6⅛ x 9¼.   0-486-40251-7

DIFFERENTIAL MANIFOLDS, Antoni A. Kosinski. Introductory text for advanced undergraduates and graduate students presents systematic study of the topological structure of smooth manifolds, starting with elements of theory and concluding with method of surgery. 1993 edition. 288pp. 5⅜ x 8½.   0-486-46244-7

VECTOR AND TENSOR ANALYSIS WITH APPLICATIONS, A. I. Borisenko and I. E. Tarapov. Concise introduction. Worked-out problems, solutions, exercises. 257pp. 5⅝ x 8¼.   0-486-63833-2

AN INTRODUCTION TO ORDINARY DIFFERENTIAL EQUATIONS, Earl A. Coddington. A thorough and systematic first course in elementary differential equations for undergraduates in mathematics and science, with many exercises and problems (with answers). Index. 304pp. 5⅜ x 8½.   0-486-65942-9

FOURIER SERIES AND ORTHOGONAL FUNCTIONS, Harry F. Davis. An incisive text combining theory and practical example to introduce Fourier series, orthogonal functions and applications of the Fourier method to boundary-value problems. 570 exercises. Answers and notes. 416pp. 5⅜ x 8½.   0-486-65973-9

COMPUTABILITY AND UNSOLVABILITY, Martin Davis. Classic graduate-level introduction to theory of computability, usually referred to as theory of recurrent functions. New preface and appendix. 288pp. 5⅜ x 8½.   0-486-61471-9

AN INTRODUCTION TO MATHEMATICAL ANALYSIS, Robert A. Rankin. Dealing chiefly with functions of a single real variable, this text by a distinguished educator introduces limits, continuity, differentiability, integration, convergence of infinite series, double series, and infinite products. 1963 edition. 624pp. 5⅜ x 8½.   0-486-46251-X

METHODS OF NUMERICAL INTEGRATION (SECOND EDITION), Philip J. Davis and Philip Rabinowitz. Requiring only a background in calculus, this text covers approximate integration over finite and infinite intervals, error analysis, approximate integration in two or more dimensions, and automatic integration. 1984 edition. 624pp. 5⅜ x 8½.   0-486-45339-1

INTRODUCTION TO LINEAR ALGEBRA AND DIFFERENTIAL EQUATIONS, John W. Dettman. Excellent text covers complex numbers, determinants, orthonormal bases, Laplace transforms, much more. Exercises with solutions. Undergraduate level. 416pp. 5⅜ x 8½.   0-486-65191-6

RIEMANN'S ZETA FUNCTION, H. M. Edwards. Superb, high-level study of landmark 1859 publication entitled "On the Number of Primes Less Than a Given Magnitude" traces developments in mathematical theory that it inspired. xiv+315pp. 5⅜ x 8½.
                                                                             0-486-41740-9

CATALOG OF DOVER BOOKS

CALCULUS OF VARIATIONS WITH APPLICATIONS, George M. Ewing. Applications-oriented introduction to variational theory develops insight and promotes understanding of specialized books, research papers. Suitable for advanced undergraduate/graduate students as primary, supplementary text. 352pp. 5⅜ x 8½.
0-486-64856-7

MATHEMATICIAN'S DELIGHT, W. W. Sawyer. "Recommended with confidence" by *The Times Literary Supplement*, this lively survey was written by a renowned teacher. It starts with arithmetic and algebra, gradually proceeding to trigonometry and calculus. 1943 edition. 240pp. 5⅜ x 8½.
0-486-46240-4

ADVANCED EUCLIDEAN GEOMETRY, Roger A. Johnson. This classic text explores the geometry of the triangle and the circle, concentrating on extensions of Euclidean theory, and examining in detail many relatively recent theorems. 1929 edition. 336pp. 5⅜ x 8½.
0-486-46237-4

COUNTEREXAMPLES IN ANALYSIS, Bernard R. Gelbaum and John M. H. Olmsted. These counterexamples deal mostly with the part of analysis known as "real variables." The first half covers the real number system, and the second half encompasses higher dimensions. 1962 edition. xxiv+198pp. 5⅜ x 8½.
0-486-42875-3

CATASTROPHE THEORY FOR SCIENTISTS AND ENGINEERS, Robert Gilmore. Advanced-level treatment describes mathematics of theory grounded in the work of Poincaré, R. Thom, other mathematicians. Also important applications to problems in mathematics, physics, chemistry and engineering. 1981 edition. References. 28 tables. 397 black-and-white illustrations. xvii + 666pp. 6⅛ x 9¼.
0-486-67539-4

COMPLEX VARIABLES: Second Edition, Robert B. Ash and W. P. Novinger. Suitable for advanced undergraduates and graduate students, this newly revised treatment covers Cauchy theorem and its applications, analytic functions, and the prime number theorem. Numerous problems and solutions. 2004 edition. 224pp. 6½ x 9¼.
0-486-46250-1

NUMERICAL METHODS FOR SCIENTISTS AND ENGINEERS, Richard Hamming. Classic text stresses frequency approach in coverage of algorithms, polynomial approximation, Fourier approximation, exponential approximation, other topics. Revised and enlarged 2nd edition. 721pp. 5⅜ x 8½.
0-486-65241-6

INTRODUCTION TO NUMERICAL ANALYSIS (2nd Edition), F. B. Hildebrand. Classic, fundamental treatment covers computation, approximation, interpolation, numerical differentiation and integration, other topics. 150 new problems. 669pp. 5⅜ x 8½.
0-486-65363-3

MARKOV PROCESSES AND POTENTIAL THEORY, Robert M. Blumental and Ronald K. Getoor. This graduate-level text explores the relationship between Markov processes and potential theory in terms of excessive functions, multiplicative functionals and subprocesses, additive functionals and their potentials, and dual processes. 1968 edition. 320pp. 5⅜ x 8½.
0-486-46263-3

ABSTRACT SETS AND FINITE ORDINALS: An Introduction to the Study of Set Theory, G. B. Keene. This text unites logical and philosophical aspects of set theory in a manner intelligible to mathematicians without training in formal logic and to logicians without a mathematical background. 1961 edition. 112pp. 5⅜ x 8½. 0-486-46249-8

INTRODUCTORY REAL ANALYSIS, A.N. Kolmogorov, S. V. Fomin. Translated by Richard A. Silverman. Self-contained, evenly paced introduction to real and functional analysis. Some 350 problems. 403pp. 5⅜ x 8½. 0-486-61226-0

APPLIED ANALYSIS, Cornelius Lanczos. Classic work on analysis and design of finite processes for approximating solution of analytical problems. Algebraic equations, matrices, harmonic analysis, quadrature methods, much more. 559pp. 5⅜ x 8½. 0-486-65656-X

AN INTRODUCTION TO ALGEBRAIC STRUCTURES, Joseph Landin. Superb self-contained text covers "abstract algebra": sets and numbers, theory of groups, theory of rings, much more. Numerous well-chosen examples, exercises. 247pp. 5⅜ x 8½.
0-486-65940-2

QUALITATIVE THEORY OF DIFFERENTIAL EQUATIONS, V. V. Nemytskii and V.V. Stepanov. Classic graduate-level text by two prominent Soviet mathematicians covers classical differential equations as well as topological dynamics and ergodic theory. Bibliographies. 523pp. 5⅜ x 8½. 0-486-65954-2

THEORY OF MATRICES, Sam Perlis. Outstanding text covering rank, nonsingularity and inverses in connection with the development of canonical matrices under the relation of equivalence, and without the intervention of determinants. Includes exercises. 237pp. 5⅜ x 8½. 0-486-66810-X

INTRODUCTION TO ANALYSIS, Maxwell Rosenlicht. Unusually clear, accessible coverage of set theory, real number system, metric spaces, continuous functions, Riemann integration, multiple integrals, more. Wide range of problems. Undergraduate level. Bibliography. 254pp. 5⅜ x 8½. 0-486-65038-3

MODERN NONLINEAR EQUATIONS, Thomas L. Saaty. Emphasizes practical solution of problems; covers seven types of equations. ". . ." a welcome contribution to the existing literature. . . ."—*Math Reviews.* 490pp. 5⅜ x 8½. 0-486-64232-1

MATRICES AND LINEAR ALGEBRA, Hans Schneider and George Phillip Barker. Basic textbook covers theory of matrices and its applications to systems of linear equations and related topics such as determinants, eigenvalues and differential equations. Numerous exercises. 432pp. 5⅜ x 8½. 0-486-66014-1

LINEAR ALGEBRA, Georgi E. Shilov. Determinants, linear spaces, matrix algebras, similar topics. For advanced undergraduates, graduates. Silverman translation. 387pp. 5⅜ x 8½. 0-486-63518-X

MATHEMATICAL METHODS OF GAME AND ECONOMIC THEORY: Revised Edition, Jean-Pierre Aubin. This text begins with optimization theory and convex analysis, followed by topics in game theory and mathematical economics, and concluding with an introduction to nonlinear analysis and control theory. 1982 edition. 656pp. 6⅛ x 9¼.
0-486-46265-X

SET THEORY AND LOGIC, Robert R. Stoll. Lucid introduction to unified theory of mathematical concepts. Set theory and logic seen as tools for conceptual understanding of real number system. 496pp. 5⅜ x 8¼. 0-486-63829-4

TENSOR CALCULUS, J.L. Synge and A. Schild. Widely used introductory text covers spaces and tensors, basic operations in Riemannian space, non-Riemannian spaces, etc. 324pp. 5⅜ x 8¼. 0-486-63612-7

ORDINARY DIFFERENTIAL EQUATIONS, Morris Tenenbaum and Harry Pollard. Exhaustive survey of ordinary differential equations for undergraduates in mathematics, engineering, science. Thorough analysis of theorems. Diagrams. Bibliography. Index. 818pp. 5⅜ x 8½. 0-486-64940-7

INTEGRAL EQUATIONS, F. G. Tricomi. Authoritative, well-written treatment of extremely useful mathematical tool with wide applications. Volterra Equations, Fredholm Equations, much more. Advanced undergraduate to graduate level. Exercises. Bibliography. 238pp. 5⅜ x 8½. 0-486-64828-1

FOURIER SERIES, Georgi P. Tolstov. Translated by Richard A. Silverman. A valuable addition to the literature on the subject, moving clearly from subject to subject and theorem to theorem. 107 problems, answers. 336pp. 5⅜ x 8½. 0-486-63317-9

INTRODUCTION TO MATHEMATICAL THINKING, Friedrich Waismann. Examinations of arithmetic, geometry, and theory of integers; rational and natural numbers; complete induction; limit and point of accumulation; remarkable curves; complex and hypercomplex numbers, more. 1959 ed. 27 figures. xii+260pp. 5⅜ x 8½. 0-486-42804-8

THE RADON TRANSFORM AND SOME OF ITS APPLICATIONS, Stanley R. Deans. Of value to mathematicians, physicists, and engineers, this excellent introduction covers both theory and applications, including a rich array of examples and literature. Revised and updated by the author. 1993 edition. 304pp. 6⅛ x 9¼. 0-486-46241-2

CALCULUS OF VARIATIONS, Robert Weinstock. Basic introduction covering isoperimetric problems, theory of elasticity, quantum mechanics, electrostatics, etc. Exercises throughout. 326pp. 5⅜ x 8½. 0-486-63069-2

THE CONTINUUM: A CRITICAL EXAMINATION OF THE FOUNDATION OF ANALYSIS, Hermann Weyl. Classic of 20th-century foundational research deals with the conceptual problem posed by the continuum. 156pp. 5⅜ x 8½. 0-486-67982-9

CHALLENGING MATHEMATICAL PROBLEMS WITH ELEMENTARY SOLUTIONS, A. M. Yaglom and I. M. Yaglom. Over 170 challenging problems on probability theory, combinatorial analysis, points and lines, topology, convex polygons, many other topics. Solutions. Total of 445pp. 5⅜ x 8½. Two-vol. set.
Vol. I: 0-486-65536-9 Vol. II: 0-486-65537-7

INTRODUCTION TO PARTIAL DIFFERENTIAL EQUATIONS WITH APPLICATIONS, E. C. Zachmanoglou and Dale W. Thoe. Essentials of partial differential equations applied to common problems in engineering and the physical sciences. Problems and answers. 416pp. 5⅜ x 8½. 0-486-65251-3

STOCHASTIC PROCESSES AND FILTERING THEORY, Andrew H. Jazwinski. This unified treatment presents material previously available only in journals, and in terms accessible to engineering students. Although theory is emphasized, it discusses numerous practical applications as well. 1970 edition. 400pp. 5⅜ x 8½. 0-486-46274-9

Math—Decision Theory, Statistics, Probability

INTRODUCTION TO PROBABILITY, John E. Freund. Featured topics include permutations and factorials, probabilities and odds, frequency interpretation, mathematical expectation, decision-making, postulates of probability, rule of elimination, much more. Exercises with some solutions. Summary. 1973 edition. 247pp. 5⅜ x 8½.
0-486-67549-1

STATISTICAL AND INDUCTIVE PROBABILITIES, Hugues Leblanc. This treatment addresses a decades-old dispute among probability theorists, asserting that both statistical and inductive probabilities may be treated as sentence-theoretic measurements, and that the latter qualify as estimates of the former. 1962 edition. 160pp. 5⅜ x 8½.
0-486-44980-7

APPLIED MULTIVARIATE ANALYSIS: Using Bayesian and Frequentist Methods of Inference, Second Edition, S. James Press. This two-part treatment deals with foundations as well as models and applications. Topics include continuous multivariate distributions; regression and analysis of variance; factor analysis and latent structure analysis; and structuring multivariate populations. 1982 edition. 692pp. 5⅜ x 8½.   0-486-44236-5

LINEAR PROGRAMMING AND ECONOMIC ANALYSIS, Robert Dorfman, Paul A. Samuelson and Robert M. Solow. First comprehensive treatment of linear programming in standard economic analysis. Game theory, modern welfare economics, Leontief input-output, more. 525pp. 5⅜ x 8½.   0-486-65491-5

PROBABILITY: AN INTRODUCTION, Samuel Goldberg. Excellent basic text covers set theory, probability theory for finite sample spaces, binomial theorem, much more. 360 problems. Bibliographies. 322pp. 5⅜ x 8½.   0-486-65252-1

GAMES AND DECISIONS: INTRODUCTION AND CRITICAL SURVEY, R. Duncan Luce and Howard Raiffa. Superb nontechnical introduction to game theory, primarily applied to social sciences. Utility theory, zero-sum games, n-person games, decision-making, much more. Bibliography. 509pp. 5⅜ x 8½.   0-486-65943-7

INTRODUCTION TO THE THEORY OF GAMES, J. C. C. McKinsey. This comprehensive overview of the mathematical theory of games illustrates applications to situations involving conflicts of interest, including economic, social, political, and military contexts. Appropriate for advanced undergraduate and graduate courses; advanced calculus a prerequisite. 1952 ed. x+372pp. 5⅜ x 8½.   0-486-42811-7

FIFTY CHALLENGING PROBLEMS IN PROBABILITY WITH SOLUTIONS, Frederick Mosteller. Remarkable puzzlers, graded in difficulty, illustrate elementary and advanced aspects of probability. Detailed solutions. 88pp. 5⅜ x 8½.   0-486-65355-2

PROBABILITY THEORY: A CONCISE COURSE, Y. A. Rozanov. Highly readable, self-contained introduction covers combination of events, dependent events, Bernoulli trials, etc. 148pp. 5⅜ x 8¼.   0-486-63544-9

THE STATISTICAL ANALYSIS OF EXPERIMENTAL DATA, John Mandel. First half of book presents fundamental mathematical definitions, concepts and facts while remaining half deals with statistics primarily as an interpretive tool. Well-written text, numerous worked examples with step-by-step presentation. Includes 116 tables. 448pp. 5⅜ x 8½.   0-486-64666-1

A TREATISE ON ELECTRICITY AND MAGNETISM, James Clerk Maxwell. Important foundation work of modern physics. Brings to final form Maxwell's theory of electromagnetism and rigorously derives his general equations of field theory. 1,084pp. $5^3/_8$ x $8^1/_2$. Two-vol. set.   Vol. I: 0-486-60636-8   Vol. II: 0-486-60637-6

MATHEMATICS FOR PHYSICISTS, Philippe Dennery and Andre Krzywicki. Superb text provides math needed to understand today's more advanced topics in physics and engineering. Theory of functions of a complex variable, linear vector spaces, much more. Problems. 1967 edition. 400pp. $6^1/_2$ x $9^1/_4$.   0-486-69193-4

INTRODUCTION TO QUANTUM MECHANICS WITH APPLICATIONS TO CHEMISTRY, Linus Pauling & E. Bright Wilson, Jr. Classic undergraduate text by Nobel Prize winner applies quantum mechanics to chemical and physical problems. Numerous tables and figures enhance the text. Chapter bibliographies. Appendices. Index. 468pp. $5^1/_8$ x $8^1/_2$.   0-486-64871-0

METHODS OF THERMODYNAMICS, Howard Reiss. Outstanding text focuses on physical technique of thermodynamics, typical problem areas of understanding, and significance and use of thermodynamic potential. 1965 edition. 238pp. $5^3/_8$ x $8^1/_2$.

      0-486-69445-3

THE ELECTROMAGNETIC FIELD, Albert Shadowitz. Comprehensive under- graduate text covers basics of electric and magnetic fields, builds up to electromagnetic theory. Also related topics, including relativity. Over 900 problems. 768pp. $5^3/_8$ x $8^1/_4$.

      0-486-65660-8

GREAT EXPERIMENTS IN PHYSICS: FIRSTHAND ACCOUNTS FROM GALILEO TO EINSTEIN, Morris H. Shamos (ed.). 25 crucial discoveries: Newton's laws of motion, Chadwick's study of the neutron, Hertz on electromagnetic waves, more. Original accounts clearly annotated. 370pp. $5^3/_8$ x $8^1/_2$.   0-486-25346-5

EINSTEIN'S LEGACY, Julian Schwinger. A Nobel Laureate relates fascinating story of Einstein and development of relativity theory in well-illustrated, nontechnical volume. Subjects include meaning of time, paradoxes of space travel, gravity and its effect on light, non-Euclidean geometry and curving of space-time, impact of radio astronomy and space-age discoveries, and more. 189 b/w illustrations. xiv+250pp. $8^3/_8$ x $9^1/_4$.   0-486-41974-6

THE VARIATIONAL PRINCIPLES OF MECHANICS, Cornelius Lanczos. Philosophic, less formalistic approach to analytical mechanics offers model of clear, scholarly exposition at graduate level with coverage of basics, calculus of variations, principle of virtual work, equations of motion, more. 418pp. $5^3/_8$ x $8^1/_2$.   0-486-65067-7

Paperbound unless otherwise indicated. Available at your book dealer, online at www.doverpublications.com, or by writing to Dept. GI, Dover Publications, Inc., 31 East 2nd Street, Mineola, NY 11501. For current price information or for free catalogues (please indicate field of interest), write to Dover Publications or log on to www.doverpublications.com and see every Dover book in print. Dover publishes more than 400 books each year on science, elementary and advanced mathematics, biology, music, art, literary history, social sciences, and other areas.